Lecture Notes in Computer Science 3075

Commenced Publication in 1973
Founding and Former Series Editors:
Gerhard Goos, Juris Hartmanis, and Jan van Leeuwen

Wolfgang Lenski (Ed.)

Logic versus Approximation

Essays Dedicated to Michael M. Richter on the Occasion of his 65th Birthday

Editor

Wolfgang Lenski
Technische Universität Kaiserslautern, Fachbereich Informatik
Postfach 30 49, 67653 Kaiserslautern, Germany
E-mail: lenski@informatik.uni-kl.de

The illustration appearing on the cover of this book is the work of Daniel Rozenberg (DADARA).

Library of Congress Control Number: 2004110441

CR Subject Classification (1998): F.4, I.2, F.2.2, G.1.2, G.2, H.2.8, D.2

ISSN 0302-9743
ISBN 3-540-22562-5 Springer Berlin Heidelberg New York

Springer is a part of Springer Science+Business Media

springeronline.com

Printed in Germany

Typesetting: Camera-ready by author, data conversion by PTP-Berlin, Protago-TeX-Production GmbH
Printed on acid-free paper SPIN: 11014485 06/3142 5 4 3 2 1 0

Preface

There are prominent moments in time where ongoing scientific work is interrupted for a moment and a more general perspective is sought. The symposion on *Logic versus Approximation* is dedicated to such a moment, namely to reflect influences of the scientific work of Michael M. Richter at the occasion of his 65th birthday. The present collection is a selection of contributions to this symposion.

In focussing on today's knowledge based systems we encounter two major paradigms of reasoning. There are on the one hand the logic-based approaches where 'logic' is to be understood in a rather broad sense. This approach is predominantly deployed in symbolic domains where numerical calculations are not the core challenge. Logic has without any doubt provided a powerful methodological tool. Progress in this area is mainly performed by refining the representation of structural aspects which results in a succession of models that should capture increasingly more aspects of the domain. This may be seen as an approximation process on the meta-level.

There is also some weakness in the logic-based approach, though, which is already due to its very foundation. The semantic theory of truth by Tarski has explicitly eliminated personal influence on the validity of truth as well as the representation of dynamically changing variations of the ground terms inside the theory. It does not allow for adaptive individual behavior per se as it is for example explicitly required in the field of e-commerce. From a methodological point of view *pragmatics* is required as opposed to *semantics*. These aspects make it worth to include rather philosophical and foundational considerations as well.

On the other hand we find approximation oriented reasoning. Methods of this kind are mainly applied in numerical domains where approximation is part of the scientific methodology itself. Here we again distinguish two different basic types, discrete and continuous domains.

However, from a more abstract level all these approaches do focus on similar topics and arise on various levels such as problem modeling, inference and problem solving mechanisms, algorithms and mathematical methods, mathematical relations between discrete and continuous properties, and are integrated in tools and applications.

Research on both kinds of reasoning in these areas has mostly been conducted independently so far. Whereas approaches based on discrete or continuous domains influence each other in a sometimes surprising way, influences between these and the symbolic approach have less intensely been studied. Especially the potentialities of an integration is certainly not understood to a satisfactory content although the primary focus from an abstract point of view is on a similar topic. It requires an unifying vision to which all parts have to contribute from their own perspective.

Scientific work is necessarily always the construction of sense. Progress is by no means arbitrary, but always guided by a quest for a still better understanding of parts of the world. Such construction processes are essentially hermeneutical ones, and an emanating coherent understanding of isolated topics is only guaranteed by a unifying view of a personal vision. Such a vision is especially provided by the research interests of Michael M. Richter that have influenced the overall perspective of the symposion. In this sense his scientific work was present all the time during the symposion.

Michael M. Richter has exerted a wide influence on logic and computer science. Although his productive work is widely spread there are some general interests behind yet. A central interest is certainly in modelling structural aspects of reality for problem solving along with the search for adequate methodologies for this purpose. Michael M. Richter made significant contributions, however, to a wide variety of topics ranging from purely logical problems in model theory and non-standard analysis to representation techniques in computer science that have finally emancipated themselves from their logical origins with special emphasis given on problems in artificial intelligence and knowledge-based systems.

The symposion on *Logic versus Approximation* has brought insight into these different approaches and contributed to the emergence of a unifying perspective. At the same time it reflected the variety of Michael M. Richter's scientific interests. The contributions to this volume range from logical problems, philosophical considerations, applications of mathematics and computer science to real world problems, programming methodologies up to current challenges in expert systems.

The members of the organization and program committee especially wish to thank the authors for submitting their papers and responding to the feedback provided by the referees. We also wish to express our gratitude to the *FAW-Förderkreis e.V.* and the *empolis GmbH* for their valuable support. Finally, we are very grateful to the local organization team of the International Conference and Research Center for Computer Science at Schloß Dagstuhl for their professionalism in handling the local arrangements.

To honor Michael M. Richter the President of the University of Kaiserslautern, Prof. Dr. Helmut J. Schmidt, in his diverting opening talk surveyed the creative powers of Michael M. Richter garnished with concise anecdotes on mutual personal experiences at Kaiserslautern University. To pay special tribute to the work of Michael M. Richter the scientific program of the symposion had been complemented by lively and inspiring after-dinner speeches by Franz Josef Radermacher und Paul Stähly. The surrondings of the International Conference Center of Schloß Dagstuhl provided the appropriate sphere and greatly helped in making the symposion a scientifically intriguing, socially enjoyable, and altogether most memorable occasion. This collection is meant to capture the essence of its scientific aspects.

Kaiserslautern, December 2003 Wolfgang Lenski

Program and Organization Committee

Althoff, Klaus-Dieter	Fraunhofer IESE Kaiserslautern
Bergmann, Ralph	Univ. Hildesheim
Lenski, Wolfgang	TU Kaiserslautern
Möhring, Rolf	TU Berlin
Radermacher, Franz Josef	FAW/Univ. Ulm

Table of Contents

A True Unprovable Formula of Fuzzy Predicate Logic 1
Petr Hájek

The Inherent Indistinguishability in Fuzzy Systems 6
Frank Klawonn, Rudolf Kruse

On Models for Quantified Boolean Formulas 18
Hans Kleine Büning, Xishun Zhao

Polynomial Algorithms for MPSP Using Parametric
Linear Programming 33
Babak Mougouie

Discrete and Continuous Methods of Demography 43
Walter Oberschelp

Computer Science between Symbolic Representation
and Open Construction 59
Britta Schinzel

Towards a Theory of Information 77
Wolfgang Lenski

Retrieval by Structure from Chemical Data Bases 106
Thomas Kämpke

Engineers Don't Search 120
Benno Stein

Randomized Search Heuristics as an Alternative
to Exact Optimization 138
Ingo Wegener

Approximation of Utility Functions by Learning Similarity Measures 150
Armin Stahl

Knowledge Sharing in Agile Software Teams 173
Thomas Chau, Frank Maurer

Logic and Approximation in Knowledge Based Systems 184
Michael M. Richter

Author Index 205

A True Unprovable Formula of Fuzzy Predicate Logic*

Petr Hájek

Institute of Computer Science
Academy of Sciences of the Czech Republic
Pod Vodárenskou věží 2, 182 07 Prague
Czech Republic

Abstract. We construct a formula true in all models of the product fuzzy predicate logic over the standard product algebra on the unit real interval but unprovable in the product fuzzy logic (and hence having truth value less than 1 in some model over a non-standard linearly ordered product algebra). Gödel's construction of a true unprovable formula of arithmetic is heavily used.

1 Introduction

Is true the same as provable? For classical logic, Gödel's completeness theorem says that a formula φ is provable in a theory T (over classical logic) iff φ is true in all models of T. On the other hand, if "true" does not mean "true in all models of the theory" but "true in the intended (standard model", Gödel's first incompleteness theorem applies: if T is a recursively (computably) axiomatized arithmetic whose axioms are true in the structure $\mathbf{N}$ of natural numbers (the standard model of arithmetic) then there is a formula true in $\mathbf{N}$ but unprovable in T. Thus such arithmetic only *approximates* the truth.

Pure classical logic (or the empty theory T with no special axioms) over classical logic does not distinguish any standard non-standard models. It has just two truth values: true and false. For fuzzy logic the situation is different since one can distinguish standard and non-standard algebras of truth functions of connectives. Fuzzy logic is a many valued logic, the standard set of truth values being the unit real interval $[0, 1]$. The truth values are ordered; one formula may be more true than another formula (comparative notion of truth). This formalizes truth of vague (imprecise) propositions (like "He is a tall man", "This is a very big number", "I shall come soon" etc.). Most often, fuzzy logic is truth-functional, i.e. works with truth functions of connectives. In particular, for basic fuzzy predicate logic $BL\forall$ (see [1]) each continuous t-norm together with its residuum determines a standard algebra of truth functions. A continuous t-norm is a continuous binary operation $*$ on the real unit square which is commutative, associative, non-decreasing in each argument, having 1 for its unit element

* Partial support of ITI (the project No. LN00A056 (ITI) of Ministry of Education (MŠMT) of the Czech Republic) is recognized.

W. Lenski (Ed.): Logic versus Approximation, LNCS 3075, pp. 1–5, 2004.

($x * 1 = x$ for all x) and having 0 for its zero element ($x * 0 = 0$ for all x). Its residuum $\Rightarrow$ is defined as follows: $x \Rightarrow y = \max\{z | x * z \leq y\}$. Any continuous t-norm can serve as the truth function of conjunction; then its residuum is taken to be the truth function of implication. This is the standard semantics of the basic fuzzy logic.

General semantics is given by the class of BL-algebras. The definition is to be found e.g. in [1], see also below Sect.2 Roughly, a BL-algebra is a (lattice)-ordered structure with a least element and a greatest element and two operations $*, \Rightarrow$ that "behave like continuous t-norms". Each continuous t-norm with its residuum defines a standard BL-algebra. Given a predicate language $\mathcal{I}$ and BL-algebra $\mathbf{L}$, an $\mathbf{L}$-interpretation $\mathbf{M}$ of $\mathcal{I}$ consists of a crisp non-empty domain M and for each predicate P of an $\mathbf{L}$-fuzzy relation (of the respective arity) on $\mathbf{M}$.[1] One gives a natural definition of the truth value $\|\varphi\|^{\mathbf{L}}_{\mathbf{M},v}$ (v being an evaluation of object variables by elements of $\mathbf{M}$) in Tarski's style by induction on the complexity of φ. For $\mathbf{L}$ being standard, $\|\varphi\|^{\mathbf{L}}_{\mathbf{M},v}$ is always defined; for general $\mathbf{L}$ one has to work with so-called safe interpretations, for which $\|\varphi\|^{\mathbf{L}}_{\mathbf{M},v}$ is total (defined for all φ). (For standard $\mathbf{L}$ each interpretation is safe.) *Axioms* and *deduction rules* of $BL\forall$ can be found in [1]; this gives the notion of *provability* in $BL\forall$. The *general completeness theorem* says that a formula φ is provable in the basic fuzzy predicate logic $BL\forall$ iff it is true (= has value 1) for each safe interpretation $\mathbf{M}$ over each linearly ordered BL-algebra $\mathbf{L}$ and each evaluation v (φ is a *general BL-tautology*).

Call φ a *standard BL-tautology* if φ is true for each (safe) interpretation over each standard BL-algebra. And as Montagna showed [4], the set of all standard BL-tautologies is not arithmetical and hence is a proper subset of the set of all general tautologies. A simple example of a formula which is a standard BL-tautology but not a general BL-tautology is the formula

$$(\forall x)(\varphi \& \nu) \equiv [(\forall x)\varphi \& \nu] \qquad (B)$$

where x is not free in ν. (A counterexample was found by F. Bou.)

Now turn to the three well-known particular continuous t-norms, namely Gödel t-norm ($x * y = \min(x, y)$), Łukasiewicz t-norm ($x * y = \max(0, x + y - 1)$) and product t-norm ($x * y = x \cdot y$) and the corresponding fuzzy predicate logics $G\forall$, Ł$\forall$, $\Pi\forall$. See [1] for details; now we only mention that the unique standard algebra of truth functions of each of those logics is given by the corresponding t-norm whereas general algebras of truth functions are algebras from the variety generated by the standard algebra, so-called MV-algebras, Gödel algebras and product algebras respectively. Axioms are those of $BL\forall$ extended by the axiom of idempotence of conjunction $\varphi \to (\varphi \& \varphi)$ for Gödel logic, the axiom of double negation ($\neg\neg\varphi \to \varphi$ where $\neg\varphi$ is $\varphi \to 0$) for Łukasiewicz logic and by two axioms presented below for product logic. All three logics have general completeness theorem (for formulas true in all safe interpretations over all linearly ordered

[1] This means that if P is n-ary then its interpretation is a mapping assigning to each n-tuple of elements of $\mathbf{M}$ an element of $\mathbf{L}$ – the degree of membership of the tuple into the interpreting fuzzy relation.

respective algebras); for Gödel logic standard tautologies coincide with general tautologies, for Łukasiewicz logic the set of standard tautologies is Π_2-complete (Ragaz), for product logic the set of standard tautologies is not arithmetical (Montagna). All three logics prove the formula (B) above (for $G\forall$ and Ł$\forall$ see [1], for $\Pi\forall$ see below). The aim of the present paper is to exhibit a particular example of a formula being a standard but not general tautology of $\Pi\forall$ (thus an unprovable standard tautology of $\Pi\forall$). As mentioned above, our formula will be a variant of Gödel's famous self-referential formula stating its own unprovability. But keep in mind that we are interested in pure (product) logic, not in a theory over this logic. To find such an example for Łukasiewicz predicate logic remains an open problem.

2 The Product Predicate Logic

Recall that a BL-algebra is a residuated lattice $\mathbf{L} = (L, \wedge, \vee, *, \Rightarrow, 0, 1)$ satisfying two additional identities

$$x \wedge y = x * (x \Rightarrow y) \quad \text{(divisibility)},$$
$$(x \Rightarrow y) \vee (y \Rightarrow x) = 1 \quad \text{(prelinearity)}.$$

Define $\neg x = x \Rightarrow 0$. A Π-algebra (product algebra) is a BL-algebra satisfying, in addition the identities

$$x \wedge \neg x = 0,$$

$$\neg\neg x \Rightarrow (((x * z) \Rightarrow (y * z)) \Rightarrow (x \Rightarrow y)) = 1.$$

The standard Π-algebra $[0,1]_\Pi$ is the unit real interval $[0,1]$ with its usual linear order ($\wedge, \vee$ being maximum and minimum), $*$ being real product and $x \Rightarrow y = 1$ for $x \leq y$, $x \Rightarrow y = y/x$ for $x > y$. $[0,1]_\Pi$ is a linearly ordered Π-algebra; each linearly ordered product algebra has Gödel negation ($\neg 0 = 1$ and $\neg x = 0$ for $x > 0$) and satisfies cancellation by a non-zero element: if $x \neq 0$ and $x * z \leq y * z$ then $x \leq y$.

Axioms of the product predicate logic $\Pi\forall$ are axioms of basic predicate fuzzy logic $BL\forall$ (see [1]) plus two additional axioms corresponding to the above identities, i.e.

$$(\varphi \wedge \neg\varphi) \to \bar{0},$$

$$\neg\neg\chi \to (((\varphi \& \chi) \to (\psi \& \chi)) \to (\varphi \to \psi)).$$

Deduction rules are modus ponens and generalization.

Recall the *general completeness:* $T \vdash \varphi$ over $\Pi\forall$ iff φ is true in each $\mathbf{L}$-model of T for each linearly ordered product $\mathbf{L}$.

Caution: In $\Pi\forall$, the quantifiers are not interdefinable; for a unary predicate U, the equivalence $(\forall x)U(x) \equiv \neg(\exists x)\neg U(x)$ is not provable. Moreover, there is a model $\mathbf{M}$ in which both $\neg(\forall x)U(x)$ and $\neg(\exists x)\neg U(x)$ are true (have value 1)

(U has a positive truth value for all objects, but the infimum of three values is 0). This was used in [2] to show that standard satisfiability in product logic is not arithmetical. Montagna improved my construction and showed that both satisfiability and tautologicity in both $\Pi\forall$ and $BL\forall$ is not arithmetical. We shall use Montagna's construction in the next section.

3 The Results

Theorem 1. *$\Pi\forall$ proves the formula* $(\forall x)(\varphi\&\nu) \equiv [(\forall x)\varphi\&\nu]$.

Proof: We give a semantic proof. It suffices to show for each linearly ordered Π-algebra $\mathbf{L}$, for $a_n \in L$ (n from an index set I) and for each $b \in L$ that

$$\inf_n (a_n * b) = (\inf_n a_n) * b.$$

Obviously, $\inf_n a_n * b \leq \inf_n (a_n * b)$. Conversely, let $(\inf_n a_n) * b < t$ for some $t \leq b$. Then $t = b * (b \Rightarrow t)$; write d for $b \Rightarrow t$; thus $(\inf_n a_n) * b < d * b$ and, by cancellation, $\inf_n a_n < d$, hence for some n, $a_n < d$, $a_n * b < d$, $\inf(a_n * b) < t$. For $t = \inf(a_n * b)$ we get a contradiction. □

In the rest of this section we shall construct a formula which is a standard tautology of $\Pi\forall$ but not a general tautology of $\Pi\forall$. We heavily use Montagna's construction from [4]. The reader is assumed to have [4] at his disposal.

P is a finite fragment of classical Peano arithmetic containing Robinson's Q, expressed for simplicity in the logic without function symbols (thus having a ternary predicate $A(x, y, x)$ for (the graph of) addition etc). Θ is the conjunction of the axioms of **P**. The language of **P** is called the arithmetical language. For each formula φ of the arithmetical language, φ° results from φ by replacing each atomic formula by its double negation. U is a new unary predicate; Ψ is the conjunction of Θ° with three axioms concerning U, namely $\neg(\forall x)U(x)$, $\neg(\exists x)\neg U(x)$ and an axiom expressing that with increasing x (in the arithmetical sense) the truth degree of $U(x)$ decreases quickly enough (Montagna's Φ_0, Φ_2, Φ_3; his Φ_1 is not necessary since we work with $\Pi\forall$). The following is the crucial fact about the construction:

Lemma 1. (1) For each model $\mathbf{M}$ of the extended language over the standard product algebra such that $\|\Psi\|^\Pi_\mathbf{M} > 0$, each closed formula φ of the arithmetical language satisfies

$$\|\varphi^\circ\|^\Pi_\mathbf{M} = 1 \text{ iff } \mathbf{N} \models \varphi$$

($\mathbf{N}$ is the standard crisp model of arithmetic).

(2) $\mathbf{N}$ has a $[0, 1]$-valued expansion $(\mathbf{N}, U^\mathbf{N})$ to a model of Ψ over the standard product algebra.

Now we apply Gödel-style diagonal lemma (cf. [3]). In **P**, arithmetize the language of $\Pi\forall$ and let $Pr_{\Pi\forall}$ be the formal provability predicate of $\Pi\forall$ expressed in **P**. Construct an arithmetical formula ν such that (**P** classically proves and hence) **N** satisfies the equivalence

$$\nu \equiv \neg Pr_{\Pi\forall}(\overline{\Psi \to \nu^\circ}).$$

(Recall that $\bar{n}$ is the n-th numeral; in **P** the above is a shorthand for the corresponding formula without function symbols.)

Lemma 2. (1) $\Pi\forall$ does not prove $\Psi \to \nu^\circ$. (2) $\mathbf{N} \models \nu$. (3) $\Psi \to \nu^\circ$ is a standard $\Pi\forall$-tautology.

Proof: (1) Assume $\Pi\forall \vdash \Psi \to \nu^\circ$; then the formulas Ψ, $\Psi \to \nu^\circ$, ν° are true in $(\mathbf{N}, U^{\mathbf{N}})$ (have the value 1), hence $\mathbf{N} \models \nu$ and thus $\Pi\forall \not\vdash \Psi \to \nu^\circ$, a contradiction.

(2) Since $\Pi\forall \not\vdash \Psi \to \nu^\circ$, $N \models \neg Pr_{\Pi\forall}(\overline{\Psi \to \nu^\circ})$ and thus $\mathbf{N} \models \nu$.

(3) Let **M** be a model of $\Pi\forall$ over the standard algebra. If $\|\Psi\|_{\mathbf{M}} = 0$ then $\|\Psi \to \nu^\circ\|_{\mathbf{M}} = 1$; if $\|\Psi\|_{\mathbf{M}} > 0$ then $\|\nu^\circ\|_{\mathbf{M}} = 1$ by Lemma 1 and hence $\|\Psi \to \nu^\circ\|_{\mathbf{M}} = 1$. □

Main theorem. *The formula $\Psi \to \nu^\circ$ is a standard $\Pi\forall$-tautology but not a general $\Pi\forall$-tautology.*

Proof: Immediate from the preceding lemma by the completeness theorem. □

References

[1] Hájek P.: Metamathematics of fuzzy logic. Kluwer 1998.

[2] Hájek P.: Fuzzy logic and arithmetical hierarchy III. Studia logica 68, 2001, pp. 135-142.

[3] Hájek P., Pudlák P.: Metamathematics of first-order arithmetic. Springer-Verlag, 1993.

[4] Montagna F.: Three complexity problems in quantified fuzzy logic. Studia logica 68, 2001, pp. 143-152.

The Inherent Indistinguishability in Fuzzy Systems

Frank Klawonn[1] and Rudolf Kruse[2]

[1] Department of Computer Science
University of Applied Sciences Braunschweig/Wolfenbüttel
Salzdahlumer Str. 46/48
D-38302 Wolfenbuettel, Germany
f.klawonn@fh-wolfenbuettel.de
[2] Department of Computer Science
Otto-von-Guericke University
Universitätsplatz 2
D-39106 Magdeburg, Germany
kruse@iik.cs.uni-magdeburg.de

Abstract. This paper provides an overview of fuzzy systems from the viewpoint of similarity relations. Similarity relations turn out to be an appealing framework in which typical concepts and techniques applied in fuzzy systems and fuzzy control can be better understood and interpreted. They can also be used to describe the indistinguishability inherent in any fuzzy system that cannot be avoided.

1 Introduction

In his seminal paper on fuzzy sets L.A. Zadeh [14] proposed to model vague concepts like *big, small, young, near, far*, that are very common in natural languages, by fuzzy sets. The fundamental idea was to allow membership degrees to sets replacing the notion of crisp membership. So the starting point of fuzzy systems is the fuzzification of the mathematical concept $\in$ (*is element of*). Therefore, a fuzzy set can be seen as generalized indicator function of a set. Where a indicator function can assume only the two values zero (standing for: is not element of the set) and one (standing for: is element of the set), fuzzy sets allow arbitrary membership degrees between zero and one.

However, when we start to fuzzify the mathematical concept of being an element of a set, it seems obvious that we might also question the idea of crisp equality and generalize it to $[0,1]$-valued equalities, in order to reflect the concept of similarity. Figure 1 shows two fuzzy sets that are almost equal. From the extensional point of view, these fuzzy sets are definitely different. But from the intensional point of view in terms of modelling vague concepts they are almost equal.

In the following we will discuss the idea of introducing the concept of (intensional) fuzzified equality (or similarity). We will review some results that on the one hand show that working with this kind of similarities leads to a better

W. Lenski (Ed.): Logic versus Approximation, LNCS 3075, pp. 6–17, 2004.

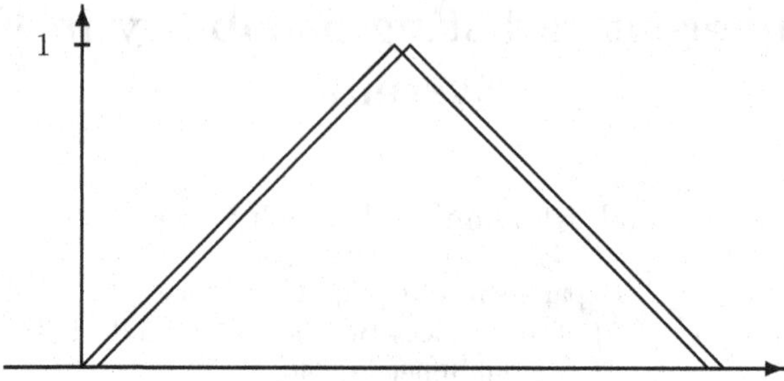

Two fuzzy sets that are almost equal.

Fig. 1. Two similar fuzzy sets

understanding of fuzzy systems and that these fuzzified equalities describe an inherent indistinguishability in fuzzy systems that cannot be overcome.

2 Fuzzy Logic

In classical logic the basics of the semantics part are truth functions for the logical connectives like $\neg, \wedge, \vee, \rightarrow, \leftrightarrow, \ldots$

Since classical logic deals with only two truth zero (false) and one (true), these truth functions can be defined in terms of simple tables as for instance for the logical connective $\wedge$ (AND):

$$\wedge : \{0,1\} \times \{0,1\} \longrightarrow \{0,1\}$$

A	B	$A \wedge B$
0	0	0
0	1	0
1	0	0
1	1	1

In the context of fuzzy sets or fuzzy systems this restriction of a two-valued logic must be relaxed to $[0,1]$-valued logic. Therefore, the truth functions of the logical connectives must be extended from the set $\{0,1\}$ to the unit interval. Typical examples for generalized truth functions $* : [0,1] \times [0,1] \longrightarrow [0,1]$ for the logical AND $\wedge$ are:

$\alpha * \beta$	name
$\min\{\alpha, \beta\}$	minimum
$\max\{\alpha + \beta - 1, 0\}$	Łuksiewicz t-norm
$\alpha \cdot \beta$	product
$\begin{cases} \min\{\alpha, \beta\} & \text{if } \max\{\alpha, \beta\} = 1 \\ 0 & \text{otherwise} \end{cases}$	drastic product

The axiomatic framework of t-norms provides a more systematic approach to extending $\wedge$ to $[0,1]$-valued logics. A **t-norm** $*$ is a commutative, associative, binary operation on [0,1] with 1 as unit that is non-decreasing in its arguments. The dual concept for the logical connective OR $\vee$ are t-conorms. A **t-conorm** $*$ is a commutative, associative, binary operation on [0,1] with 0 as unit that is non-decreasing in its arguments. A t-norm $*$ induces a t-conorm $\tilde{*}$ by $\alpha \tilde{*} \beta = 1 - ((1-\alpha) * (1-\beta))$ and vice versa.

In this paper, we will restrict our consideration to continuous t-norms. In this case, we can introduce the concept of residuated implication. $\rightarrow_*$ is called **residuated implication** w.r.t. $*$, if

$$\alpha * \beta \leq \gamma \Rightarrow \alpha \leq \beta \rightarrow_* \gamma$$

holds for all $\alpha, \beta, \gamma \in [0,1]$.

A continuous t-norm $*$ has a unique residuated implication given by

$$\alpha \rightarrow_* \beta \;=\; \bigvee\{\lambda \in [0,1] \mid \alpha * \lambda \leq \beta\}.$$

The **biimplication** w.r.t. to the (residuated) implication $\rightarrow_*$ is defined by

$$\alpha \leftrightarrow_* \beta \;=\; (\alpha \rightarrow_* \beta) \wedge (\beta \rightarrow_* \alpha).$$

The **negation** w.r.t. to the (residuated) implication $\rightarrow_*$ is defined by

$$\neg_* \alpha \;=\; \alpha \rightarrow_* 0.$$

The most common examples for t-norms and induced logical connectives are:

1. $\alpha * \beta \;=\; \min\{\alpha, \beta\}$

$$\alpha \rightarrow \beta = \begin{cases} 1 & \text{if } \alpha \leq \beta \\ \beta & \text{otherwise} \end{cases}$$

$$\alpha \leftrightarrow = \begin{cases} 1 & \text{if } \alpha = \beta \\ \min\{\alpha, \beta\} & \text{otherwise} \end{cases}$$

$$\neg\alpha = \begin{cases} 1 & \text{if } \alpha = 0 \\ 0 & \text{otherwise} \end{cases}$$

2. $\alpha * \beta \;=\; \max\{\alpha + \beta - 1, 0\}$

$$\alpha \rightarrow \beta = \min\{1 - \alpha + \beta, 1\}$$
$$\alpha \leftrightarrow \beta = 1 - |\alpha - \beta|$$
$$\neg\alpha = 1 - \alpha$$

3. $\alpha * \beta \;=\; \alpha \cdot \beta$

$$\alpha \rightarrow \beta = \begin{cases} 1 & \text{if } \alpha \leq \beta \\ \frac{\beta}{\alpha} & \text{otherwise} \end{cases}$$

$$\alpha \leftrightarrow \beta) = \begin{cases} 1 & \text{if } \alpha = \beta \\ \frac{\min\{\alpha,\beta\}}{\max\{\alpha,\beta\}} & \text{otherwise} \end{cases}$$

$$\neg\alpha = \begin{cases} 1 & \text{if } \alpha = 0 \\ 0 & \text{otherwise} \end{cases}$$

If $[\![A]\!]$ denotes the truth value of the logical formula A, then the truth functions for quantifiers are given by

$$[\![(\forall x)(A(x))]\!] = \bigwedge_x [\![A(x))]\!] \quad \text{and} \quad [\![(\exists x)(A(x))]\!] = \bigvee_x [\![A(x))]\!]$$

It should be mentioned that these concepts of $[0, 1]$-valued logics lead to interesting generalizations of classical logic from the purely mathematical point of view. However, the assumption of truth-functionality, i.e. that the truth value of a complex logical formula depends only on the truth values of its compound elements, leads to certain problems. Truth-functionality implies a certain independence assumption between the logical formulae. Like in probability theory, independence is a very strong assumption that is seldom satisfied in practical applications.

Already the simple example of three-valued Łukasiewicz logic illustrates this problems. The third truth value u in this logic stands for *undetermined*. The logical connective $\wedge$ is defined canonically by the following truth function:

$$* : \{0, u, 1\} \times \{0, u, 1\} \longrightarrow \{0, u, 1\}$$

A	B	$A * B$
0	0	0
0	u	0
0	1	0
u	0	0
u	u	u
u	1	u
1	0	0
1	u	u
1	1	1

The following simple example shows the problem caused by truth functionality.

Proposition	Meaning	$[\![\ldots]\!]$
A	The German chancellor will be in Berlin on 30 November 2010.	u
B	It will rain in Berlin on 30 November 2010.	u
$A \wedge B$	...	u
$A \wedge \neg A$	...	0

A and B are independent (hopefully). A and $\neg A$ are definitely not. So there is no consistent way of assigning a truth value to a logical conjunction of two statements based only on their truth values, since A, B and $\neg A$ all have the truth value *undetermined*, as well as the logical statement $A \wedge B$, whereas $A \wedge \neg A$ should be assigned the truth value false.

However, in applications of fuzzy systems like fuzzy control this problem usually plays only a minor role, because certain independence assumptions are satisfied there by the structure of the considered formal framework.

3 Similarity Relations

Before introducing the notion fuzzified equality or similarity, we briefly review how mathematical concepts can be fuzzified in a straight forward way.

We interpret the membership degree $\mu(x)$ of an element x to a fuzzy set μ as the truth value of the statement *x is an element of μ*.

$$\mu(x) \;=\; [\![x \in \mu]\!]$$

When we want to consider the fuzzified version of an axiom A (in classical logic), we take into account that axioms are assumed to be true, i.e. $[\![A]\!] = 1$. Also, axioms are very often of the form $B \to C$. Using residuated implications, we have

$$[\![B \to C]\!] = 1 \;\Leftrightarrow\; [\![B]\!] \leq [\![C]\!]$$

Having these facts in mind, it is obvious how to interpret an axiom in a $[0,1]$-valued logic. As a concrete example, we consider the notion of equivalence relations.

classical logic	fuzzy logic
relation: $E \subseteq X \times X$	fuzzy relation: $E : X \times X \longrightarrow [0,1]$
$(x,x) \in E$	$E(x,x) = 1$
$(x,y) \in E \Rightarrow (y,x) \in E$	$E(x,y) \leq E(y,x)$ (thus $E(x,y) = E(y,x)$)
$(x,y) \in E \wedge (y,z) \in E \Rightarrow (x,z) \in E$	$E(x,y) * E(y,z) \leq E(x,z)$

A fuzzy relation

$$E : X \times X \longrightarrow [0,1]$$

on a set X satisfying the three previously mentioned axioms is called an **similarity relation** [15,11]. Depending on the choice of the operation $*$, sometimes E is also called an indistinguishability operator [13], fuzzy equality (relation) [2, 7], fuzzy equivalence relation [12] or proximity relation [1].

A fuzzy relation E is a similarity relation w.r.t. the Łukasiewicz t-norm, if and only if $1 - E$ is a pseudo-metric bounded by 1. A fuzzy relation E is an similarity relation w.r.t. the minimum, if and only if $1 - E$ is an ultra-pseudo-metric bounded by 1. Any (ultra-)pseudo-metric δ bounded by 1 induces an similarity relation w.r.t. the Łukasiewicz t-norm (minimum) by $E = 1 - \delta$.

Extensionality in the context of similarity relation means to respect the similarity relation: Equal (similar) elements should lead to equal (similar) results. The classical property: $x \in M \wedge x = y \;\;\Rightarrow\;\; y \in M$ leads to the following definition.

A fuzzy set μ is called **extensional** w.r.t. an similarity relation E, if

$$\mu(x) * E(x, y) \leq \mu(y)$$

holds.

Let $E : X \times X \to [0, 1]$ be an similarity relation on the set X. The extensional hull $\hat{\mu}$ of the fuzzy set μ is smallest extensional fuzzy set containing μ given by

$$\hat{\mu}(y) = \bigvee_{x \in X} (\mu(x) * E(x, y)).$$

$$y \in \mu \Leftrightarrow (\exists x \in X)(x \in \mu \wedge x = y)$$

The extensional hull of an ordinary set is the extensional hull of its indicator function and can be understood as the (fuzzy) set of points that are equal to at least one element in the set.

As an example consider the similarity relation $E(x, y) = 1 - \min\{|x - y|, 1\}$. The extensional hulls of a single point and an interval are shown in figure 2.

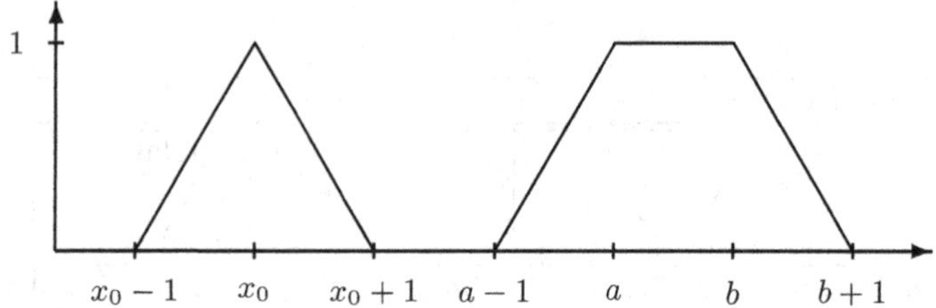

Fig. 2. The extensional hulls of the point x_0 and of the interval $[a, b]$.

Extensional hulls of points w.r.t.

$$E(x, y) = 1 - \min\{|x - y|, 1\}$$

always have a support of length two. In order to maintain the degrees of similarity (and the membership degrees), when changing the measurement unit (seconds instead of hours, miles instead of kilometers,...), we have to take a scaling into account:

$$E(x, y) = 1 - \min\{c \cdot |x - y|, 1\}$$

When we take a closer look at the concept of similarity relations, we can even introduce a more general concept of scaling. Similarity relations can be used to model indistinguishability. There are two kinds of indistinguishability, we have to deal with in typical fuzzy control applications.

Enforced indistinguishability is caused by limited precision of measurement instruments, (imprecise) indirect measurements, noisy data, ...

Intended indistinguishability means that the control expert is not interested in more precise values, since a higher precision would not really lead to an improved control.

Both kinds of indistinguishability might need a local scaling as the following example of designing an air conditioning system shows.

temperature (in °C)	scaling factor	interpretation
< 15	0.00	exact value meaningless (much too cold)
15-19	0.25	too cold, but not too far away from the desired temperature, regulation need not be too sensitive
19-23	1.50	very sensitive, near the optimal value
23-27	0.25	too warm, but not too far away from the desired temperature, regulation need not be too sensitive
> 27	0.00	exact value meaningless (much too hot)

When we apply these different scaling factors to our temperature domain, this has the following consequences, when we consider the similarity relation induced by the scaled distance. In order to determine how dissimilar two temperatures are, we do note compute their difference directly, but in the scaled domain, where the range up to 15 is shrunk to a single point, the range between 15 and 19 is shrunk by the factor 0.25, the range between 19 and 23 is stretched by the factor 1.5 and so on. The following table shows the scaled distances of some example values for the temperature.

pair of values (x, y)	scal. factor c	transformed distance $\delta(x,y) = \|c \cdot x - c \cdot y\|$	similarity degree $E(x,y) = 1 - \min\{\delta(x,y), 1\}$
(13,14)	0.00	0.000	1.000
(14,14.5)	0.00	0.000	1.000
(17,17.5)	0.25	0.125	0.875
(20,20.5)	1.50	0.750	0.250
(21,22)	1.50	1.500	0.000
(24,24.5)	0.25	0.125	0.875
(28,28.5)	0.00	0.000	1.000

Figures 3 and 4 show examples of extensional hulls of single points.

The idea of piecewise constant scaling functions can be extended to arbitrary scaling functions in the following way [4]. Given an integrable scaling function: $c : R \to [0, \infty)$. If we assume that we have for small values ε that the transformed distance between x and $x + \varepsilon$ is given by

$$\delta(x, x+\varepsilon) \approx c \cdot |\varepsilon|,$$

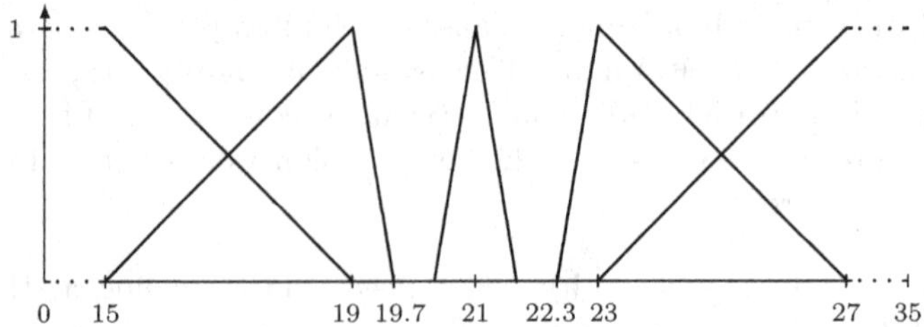

Fig. 3. The extensional hulls of the points 15, 19, 21, 23 and 27.

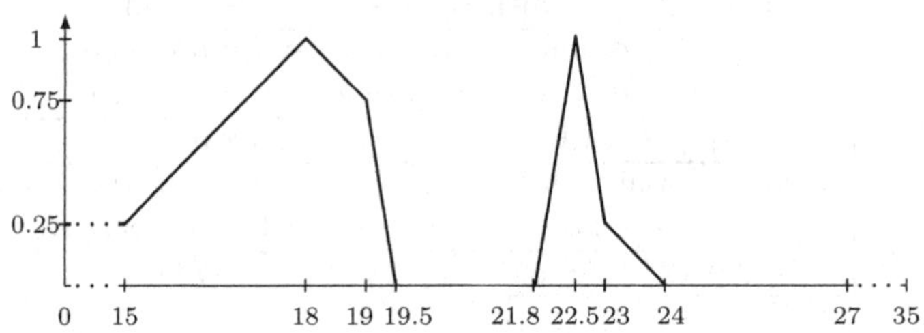

Fig. 4. The extensional hulls of the points 18.5 and 22.5.

then the transformed distance induced by the scaling function c can be computed by

$$\left| \int_x^y c(s)\, ds \right| .$$

4 The Inherent Indistinguishability in Fuzzy Systems

In this section we present some results [4,6,5] on the connection between fuzzy sets and similarity relations.

Given a set $\mathcal{A}$ of fuzzy sets ('a fuzzy partition'). Is there an similarity relation E s.t. all these fuzzy sets are extensional w.r.t. E? The answer to this question is positive.

$$E_{\mathcal{A}}(x,y) \;=\; \bigwedge_{\mu\in\mathcal{A}} (\mu(x) \leftrightarrow \mu(y))$$

is the coarsest similarity relation making all fuzzy sets in $\mathcal{A}$ extensional.

We go a step further and consider a given set $\mathcal{A}$ of normal fuzzy sets (that have membership degree one for at least one point). Is there an similarity relation E s.t. all these fuzzy sets can be interpreted as extensional hulls of points?

Let $\mathcal{A}$ be a set of fuzzy sets such that for each $\mu \in \mathcal{A}$ there exists $x_\mu \in X$ with $\mu(x_\mu) = 1$. There is an similarity relation E, such that for all $\mu \in \mathcal{A}$ the extensional hull of the point x_μ coincides with the fuzzy set μ, if and only if

$$\bigvee_{x \in X} (\mu(x) * \nu(x)) \leq \bigwedge_{y \in X} (\mu(y) \leftrightarrow \nu(y))$$

holds for all $\mu, \nu \in \mathcal{A}$.

In this case, $E = E_{\mathcal{A}}$ is the coarsest similarity relation for which the fuzzy sets in $\mathcal{A}$ can be interpreted as extensional hulls of points.

If the fuzzy sets are pairwise disjoint ($\mu(x) * \nu(x) = 0$ for all x), then the condition of the previous theorem is always satisfied. For the Łukasiewicz t-norm this means

$$\mu(x) + \nu(x) \leq 1.$$

Let $\mathcal{A}$ be a non-empty, at most countable set of fuzzy sets such that each $\mu \in \mathcal{A}$ satisfies:

- There exists $x_\mu \in R$ with $\mu(x_\mu) = 1$.
- μ (as a real-valued function) is increasing on $(-\infty, x_\mu]$.
- μ is decreasing on $[x_\mu, -\infty)$.
- μ is continuous.
- μ is differentiable almost everywhere.

There exists a scaling function $c : R \to [0, \infty)$ such that for all $\mu \in \mathcal{A}$ the extensional hull of the point x_μ w.r.t. the similarity relation

$$E(x, y) = 1 - \min\left\{ \left| \int_x^y c(s)\, ds \right|, 1 \right\}$$

coincides with the fuzzy set μ, if and only if

$$\min\{\mu(x), \nu(x)\} > 0 \Rightarrow \left| \frac{d\mu(x)}{dx} \right| = \left| \frac{d\nu(x)}{dx} \right|$$

holds for all $\mu, \nu \in \mathcal{A}$ almost everywhere. In this case,

$$c : R \to [0, \infty), \quad x \mapsto \begin{cases} \left| \frac{d\mu(x)}{dx} \right| & \text{if } \mu(x) > 0 \\ 0 & \text{otherwise} \end{cases}$$

can be chosen as the (almost everywhere well-defined) scaling function.

Figure 5 shows a typical example of a choice of fuzzy sets. For this kind of fuzzy partition a scaling function exists, such that the fuzzy sets can be represented as extensional hulls of points.

There is another explanation, why fuzzy sets are very often chosen as shown in this figure. The expert who specifies the fuzzy sets and the rules for the fuzzy system is assumed to specify as few rules as possible. When he has chosen one point (inducing a fuzzy set as its extensional hull), taking the similarity relations into account, this single point provides some information for all points that have non-zero similarity/indistinguishability to the specified point. Therefore, the next point must be specified, when the similarity degree (membership degree of the corresponding fuzzy set) has dropped to zero.

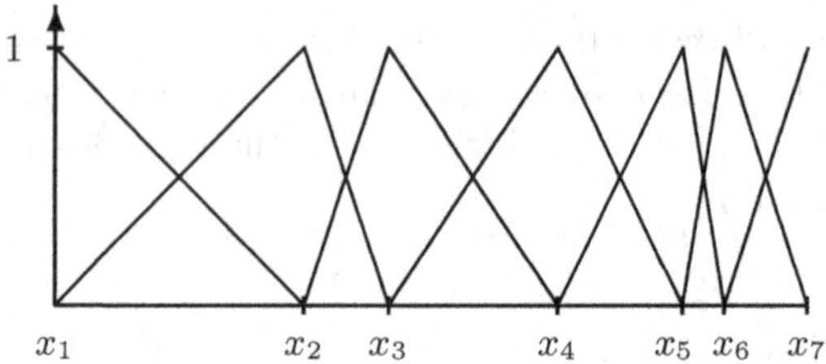

Fig. 5. Fuzzy sets for which a scaling function can be defined.

5 Similarity Relations and Fuzzy Functions

If we assume that the fuzzy sets in fuzzy control applications represent (vague) points, then each rule specifies a point on the graph of the control function. A rule is typically of the form

If x_1 is μ_1 and ... and x_n is μ_n, then y is ν.

where $x_1, \ldots, x_n$ are input variables and y is the output variable and $\mu_1, \ldots, \mu_n$ and ν are suitable fuzzy sets.

In this way, fuzzy control can be seen as interpolation in the presence of vague environments characterized by similarity relations. A function $f : X \longrightarrow Y$ is extensional w.r.t. to the similarity relations E and F on X and Y, respectively, if

$$E(x, x') \leq F(f(x), f(x'))$$

holds for all $x, x' \in X$.

Interpreting fuzzy control in this way, defuzzification means to find an extensional function that passes through the points specified by the rule base. It can be shown [10] that the centre of gravity defuzzification method is a reasonable heuristic technique, when the fuzzy sets and the rules are 'well-behaved'. From a theoretical point of view, we have to find a function through the given control points that is Lipschitz continuous (w.r.t. the metrics induced by the equality relations) with Lipschitz constant 1.

Since fuzzy controllers usually have multiple inputs, it is necessary to combine the similarity relations to a single similarity relation in the product space. The canonical similarity relation on a product space is given by [9]

$$E((x_1, \ldots, x_p), (x'_1, \ldots, x'_p)) = \min_{i \in \{1, \ldots, p\}} \{E_i(x_i, x'_i)\}.$$

In terms of fuzzy control this means that for a single rule, the membership degrees of an input would be combined using the minimum.

Viewing fuzzy control in this way, the specification of (independent) fuzzy sets respectively similarity relations means that the indistinguishabilities on the different inputs are independent. Although this is an unrealistic assumption, fuzzy control works quite well. The independence problem is partly solved, by using a fine granularity everywhere and specifying more rules.

Finally, we would like to emphasize that, even if the fuzzy sets are chosen in such a way that they cannot be interpreted as extensional hulls of points, similarity relations play an important role. We can always compute the coarsest similarity relations making all fuzzy sets extensional. It can be shown under quite general assumptions [6] that

- the output of a fuzzy system does not change, when we replace the input by its extensional hull and
- the output (before defuzzification) is always extensional.

6 Conclusions

We have shown that similarity relations provide an interesting framework to better understand the concepts underlying fuzzy systems and fuzzy control. They can also be used to characterize the indistinguishability that is inherent in any fuzzy system. Exploiting the ideas of the connection between fuzzy systems and similarity relations further leads also to interesting connections to fuzzy clustering [3] and to understanding fuzzy control as knowledge-based interpolation [8] which leads to a much stricter framework of fuzzy systems in which inconsistencies can be avoided easier [5]

References

1. Dubois, D., Prade, H.: Similarity-Based Approximate Reasoning. In: Zurada, J.M., Marks II, R.J., Robinson, C.J. (eds.): Computational Intelligence Imitating Life. IEEE Press, New York (1994), 69-80
2. Höhle, U., Stout, L.N.: Foundations of Fuzzy Sets. Fuzzy Sets and Systems 40 (1991), 257-296
3. Höppner, F., Klawonn, F., Kruse, R., Runkler. T.: Fuzzy Cluster Analysis. Wiley, Chichester (1999)
4. Klawonn, F.:Fuzzy Sets and Vague Environments. Fuzzy Sets and Systems 66 (1994), 207-221
5. Klawonn, F.: Fuzzy Points, Fuzzy Relations and Fuzzy Functions. In: Novák, V., Perfilieva, I. (eds.): Discovering the World with Fuzzy Logic. Physica-Verlag, Heidelberg (2000), 431-453
6. Klawonn, F., Castro, J.L.: Similarity in Fuzzy Reasoning. Mathware and Soft Computing 2 (1995), 197-228
7. Klawonn, F., Kruse, R.:Equality Relations as a Basis for Fuzzy Control. Fuzzy Sets and Systems 54 (1993), 147-156
8. Klawonn, F., Gebhardt, J., Kruse, R.: Fuzzy Control on the Basis of Equality Relations – with an Example from Idle Speed Control. IEEE Transactions on Fuzzy Systems 3 (1995), 336-350
9. Klawonn, F., Novák, V.: The Relation between Inference and Interpolation in the Framework of Fuzzy Systems. Fuzzy Sets and Systems 81 (1996), 331-354

10. Kruse, R., Gebhardt, J., Klawonn, F.: Foundations of Fuzzy Systems. Wiley, Chichester (1994)
11. Ruspini, E.H.: On the Semantics of Fuzzy Logic. Intern. Journ. of Approximate Reasoning 5 (1991), 45-88
12. Thiele, H., Schmechel, N.: The Mutual Defineability of Fuzzy Equivalence Relations and Fuzzy Partitions. Proc. Intern. Joint Conference of the Fourth IEEE International Conference on Fuzzy Systems and the Second International Fuzzy Engineering Symposium, Yokohama (1995), 1383-1390
13. Trillas, E., Valverde, L.: An Inquiry into Indistinguishability Operators. In: Skala, H.J., Termini, S., Trillas, E. (eds.): Aspects of Vagueness. Reidel, Dordrecht (1984), 231-256
14. Zadeh, L.A: Fuzzy Sets. Information and Control 8 (1965), 338-353.
15. Zadeh, L.A.: Similarity Relations and Fuzzy Orderings. Information Sciences 3 (1971), 177-200

On Models for Quantified Boolean Formulas

Hans Kleine Büning[1] and Xishun Zhao[2]

[1] Department of Computer Science, Universität Paderborn
33095 Paderborn (Germany)
kbcsl@upb.de

[2] Institute of Logic and Cognition, Zhongshan University
510275, Guangzhou, (P.R. China)
hsdp08@zsu.edu.cn

Abstract. A quantified Boolean formula is true, if for any existentially quantified variable there exists a Boolean function depending on the preceding universal variables, such that substituting the existential variables by the Boolean functions results in a true formula. We call a satisfying set of Boolean functions a model. In this paper, we investigate for various classes of quantified Boolean formulas and various classes of Boolean functions the problem whether a model exists. Furthermore, for these classes the complexity of the model checking problem - whether a set of Boolean functions is a model for a formula - will be shown. Finally, for classes of Boolean functions we establish some characterizations in terms of quantified Boolean formulas which have such a model. For example, roughly speaking any satisfiable quantified Boolean Horn formula can be satisfied by monomials and vice versa.

Keywords: quantified Boolean formula, Boolean function, model checking, complexity, satisfiability

1 Introduction

Quantified Boolean formulas (*QBF*) are a powerful tool for the representation of many problems in computer science and artificial intelligence such as planning, abductive reasoning, non-monotonic reasoning, intuitionistic and modal logics. This has motivated to design efficient decision algorithms for *QBF* or to look for tractable subclasses of *QBF* which are still able to formulate some interesting problems in practice.

The most natural approach to determine the truth of a *QBF* formula is based on the semantics. In other words, this procedure iteratively splits the formula $Qx\Phi$ of the problem into two simpler formulas $\Phi[x = 1]$ and $\Phi[x = 0]$. Some improvements for this approach by using various backtracking strategies have been made [2,4,6,9].

Another method for solving the satisfiability problem for *QBF*is the Q-resolution which is the generalization of the resolution approach for propositional formulas [7,5].

W. Lenski (Ed.): Logic versus Approximation, LNCS 3075, pp. 18–32, 2004.

Quantified Boolean formulas can be used to describe games. Let us consider a simple game. $\Phi = \forall x_1 \exists y_1 \cdots \forall x_n \exists y_n G(x_1, y_1, \cdots, x_n, y_n)$ may describe a two-person game, where x_i is the i–the move of the first player and y_j is the j–th move of the second player. The moves are 0 or 1. The propositional formula $G(x_1, y_1, \cdots, x_n, y_n)$ may represent the situation that for the moves $x_1, y_1, \cdots, x_n, y_n$ player 2 wins the game.

Now several questions arise: The first question is whether there exists a winning strategy for player 2. That means, we have to decide whether the quantified Boolean formula is true. Φ is true, if there is an assignment of truth values to the existentially quantified variables depending on the preceding universally quantified variables, such that the formula $G(x_1, y_1, \cdots, x_n, y_n)$ is true or in other words a tautology. The assignment to the existentially quantified variables can be expressed in terms of Boolean functions. We denote the set of satisfying Boolean functions as a model. The term model is chosen, because a propositional formula ϕ over the variables $x_1, \cdots, x_n$ is satisfiable if and only if $\exists x_1 \cdots \exists x_n \phi$ is true. Often, a satisfying truth assignment for ϕ is called a model. For formulas in first order logic, a model for a formula assigns besides the assignments for predicates, the domain etc. functions to function symbols and via skolemization functions to existential variables. In terms of games or other applications one may replace the term model by strategies, policies, or plan of actions.

In general, the problem whether a quantified Boolean formula has a model is equivalent to the satisfiability problem for quantified Boolean formulas and therefore *PSPACE*–complete [10].

With respect to the game, now the question is, whether there is a sequence of moves, or in other words Boolean functions, which may fulfill further requirements or belong to a fixed class of Boolean functions. That leads to the problem whether for a fixed class of Boolean functions there is a model for the formula within this class.

Suppose, we have a strategy in mind, say the Boolean functions $f_1, \cdots, f_n$. We want to see whether this set of functions is a model for Φ. This is denoted as the model checking problem. More general, we have to decide whether for a given formula Φ, a set of Boolean functions M is a model.

The last question is the following: Suppose, we take Φ from a subclass of quantified Boolean formulas, for example quantified Horn formulas *QHORN*. Can we always find simple models, say for example Boolean functions represented as monomials? For *QHORN* we know that for any satisfiable formula a model in terms of monomials can be found.

The other way round, for a given class of Boolean functions K we want to determine the set of quantified Boolean formulas which can be satisfied by models consisting of functions in K.

Using Boolean functions as models may lead to some confusion, because the functions can be represented by truth tables, propositional formulas or quantified Boolean formulas with free variables. The representation may have an essential effect on the complexity of the various problems. In general, truth tables require more space than propositional formulas and propositional formulas more space than quantified Boolean formulas with free variables. Subsequently, up to the last subsection, we demand that Boolean functions are given as propositional formulas.

Without any restrictions to the Boolean functions and to the quantified Boolean formulas the model problem is *PSPACE*-complete, whereas we will see that the model checking problem is *coNP*-complete. An overview of the main results will be presented at the beginning of Section 5.

2 Quantified Boolean Formulas

In this section we recall the basic notions for propositional and quantified Boolean formulas and introduce some terminology.

A literal is a propositional variable or a negated variable. Clauses are disjunctions of literals. A *CNF* formula is a conjunction of clauses. For *CNF* formulas with clause–length less or equal than k, we write *k-CNF*. *HORN* is the set of Horn formulas. That means, any clause contains at most one non–negated literal. *DNF* is the set of propositional formulas in disjunctive normal form.

A *QBF* formula Φ is a quantified Boolean formula *without free variables*. The formula Φ is in prenex normal form, if $\Phi = Q_1x_1 \cdots Q_nx_n\phi$, where $Q_i \in \{\forall, \exists\}$ and ϕ is a propositional formula over variables $x_1, \cdots, x_n$. $Q_1x_1 \cdots Q_nx_n$ is called the prefix and ϕ the matrix of Φ. Usually, we simply write $\Phi = Q\phi$.

A literal x or $\neg x$ is called a universal resp. existential literal, if the variable x is bounded by a universal quantifier resp. by an existential quantifier. Universally resp. existentially quantified variables are also denoted as $\forall$-variables resp. $\exists$-variables. A clause containing only $\forall$-literals is called universal clause or $\forall$-disjunction.

QCNF denotes the class of *QBF* formulas with matrix in *CNF*.

Definition 1. *Let* $\Phi = Q(\alpha_1 \wedge \cdots \wedge \alpha_n)$ *and* $\Phi' = Q(\alpha'_i \wedge \cdots \wedge \alpha'_n)$ *be two formulas in QCNF. If* α'_i *is a subclause of* α_i *for every* i, *then we write* $\Phi' \subseteq_{cl} \Phi$. *We say* Φ' *is a* subclause–subformula *of* Φ.

For example, we have $\forall x \forall z \exists y (z \vee y) \wedge (\neg x \vee \neg y) \subseteq_{cl} \forall x \forall z \exists y (x \vee z \vee y) \wedge (\neg x \vee \neg y)$.

A closed formula $\Phi \in QCNF$ is called satisfiable or true, if there exists an assignment of truth values to the existential variables depending on the preceding universal variables, such that the propositional kernel of the formula is true. For $\Phi = \forall x \exists y (x \vee y) \wedge (\neg x \vee \neg y)$ choosing for the value of y the value of $\neg x$ the formula $(\neg x \vee y) \wedge (x \vee \neg y)$ is true. The assignment can be considered as a function $f(x) = \neg x$.

For quantified Boolean formulas with free variables, denoted as *QBF**, the formula is satisfiable if, and only if there is a truth assignment for the free variables, such that for the truth assignment the closed *QBF* is true. We write $\Phi(x_1, \cdots, x_n) = Q\phi(x_1, \cdots, x_n)$ for a *QCNF**, if Q is the prefix, ϕ is the kernel, and $x_1, \cdots, x_n$ are the free variables. For example, the formula $\Phi(x) = \forall z \exists y (z \vee$

$y) \wedge (x \vee \neg y)$ has the free variable x and is satisfiable, because for $x = 1$ we obtain the satisfiable formula $\Phi' = \forall z \exists y(z \vee y)$.

In general, we assume that any *QCNF* input formula contains no tautological clause. With respect to the satisfiability, in formulas of the form $\Phi = Q\forall x\phi$ we can remove the universal quantification $\forall x$ in the prefix and delete all occurrences of x preserving the satisfiability, if x is not a unit clause in ϕ. In that case Φ is not true.

In our investigations we make use of substitutions of variables by formulas. For a formula $\Phi(x_1, \cdots, x_n)$ with or without free variables $\Phi(x_1 \cdots, x_n)[y_1/f_1, \cdots, y_n/f_n]$ denotes the formula obtained by simultaniously substituting the occurrences of the variables y_i by the formula f_i. For example, $\Phi(z)[y/\neg x] = \forall x \exists y(z \vee x \vee y) \wedge (\neg x \vee \neg y)[y/\neg x]$ results in the formula $\Phi'(z) = \forall x \exists y(z \vee x \vee \neg x) \wedge (\neg x \vee x)$. In $\Phi'(z)$ we can delete the existential variable y in the prefix.

For a formula $\Phi = Q\phi$ in *QCNF*, $\phi_{|\exists}$ is the propositional formula obtained from ϕ by removing every $\forall$-literal.

For a class $\mathcal{X}$ of propositional formulas $Q\mathcal{X}$ denotes the class of quantified Boolean formulas in prefix normal form with matrix in $\mathcal{X}$. For example, *QHORN* is the class of quantified Boolean Horn formulas.

3 Boolean Models

In this section we give a formal definition of models for quantified Boolean formulas without free variables and introduce the model and the model checking problem. In the remainder we show some basic results dealing with the complexity of these problems.

Definition 2. *(Model)*
Let $\Phi = \forall z_1 \exists y_1 \cdots \forall z_m \exists y_m \phi$ be a satisfiable formula in QBF. For $(1 \leq i \leq m)$ and Boolean functions $f_{y_i}(z_1, \cdots z_i)$ we say $M = (f_{y_1}, \cdots, f_{y_m})$ is a model for Φ if and only if
$\forall z_1 \cdots \forall z_t \phi[y_1/f_{y_1}(z_1), \cdots, y_m/f_{y_m}(z_1, \cdots, z_m)]$ is true.
If the Boolean functions f_i belong to a class K of Boolean functions, then we say M is a K–model for Φ.

Example 1. :
The formula $\Phi = \forall x \exists y(x \vee y) \wedge (\neg x \vee \neg y)$ is true and for $f_y(x) = \neg x$, $M = (f_y)$ is a model, because
$\forall x(x \vee y) \wedge (\neg x \vee \neg y)[y/f_y(x)] = \forall x(x \vee f_y(x)) \wedge (\neg x \vee \neg f_y(x)) = \forall x(x \vee \neg x) \wedge (\neg x \vee x)$ contains tautological clauses only.

When not stated otherwise, we represent Boolean functions by propositional formulas. Only in the last subsection we consider another representation, namely quantified Boolean formulas with free variables..

Let K be a class of propositional formulas and $X \subseteq QCNF$.

K–Model Checking Problem for X:
Instance: A formula $\Phi \in X$ and $M = (f_1, \cdots, f_n)$ a sequence of propositional formulas $f_i \in K$.
Query: Is M a K–model of Φ?

K–Model Problem for X:
Instance: A formula $\Phi \in X$.
Query: Does there exist a K–model M for Φ?

The *CNF*–model problem for *QCNF* is *PSPACE*–complete, since any Boolean function can be represented as a *CNF* formula. Therefore, to decide whether a formula Φ has a model is equivalent to the question whether Φ is satisfiable. That problem is known to be *PSPACE*–complete [7,10].

Lemma 1. *The CNF–model checking problem for QCNF is coNP–complete.*

Proof: Let $\Phi = \forall x_1 \exists y_1 \cdots \forall x_n \exists y_n \phi$ be a formula in *QCNF* and $M = (f_1, \cdots, f_n)$ be a sequence of propositional formulas for Φ. Then M is a model for Φ i.e. $\Phi' = \forall x_1 \cdots \forall x_n \phi[y_1/f_1(x_1), \cdots, y_n/f_n(x_1, \cdots, x_n)]$ is true if and only if the propositional formula $\phi[y_1/f_1(x_1), \cdots, y_n/f_n(x_1, \cdots, x_n)]$ is a tautology. Since the tautology problem for propositional formulas is *coNP* –complete and $|\Phi'| \leq |\Phi||M|$, the model checking problem is in *coNP*.
The *coNP*–hardness can be shown as follows: For $\Phi = \forall x_1 \cdots \forall x_n \exists y(y)$ and an arbitrary propositional formula f over the variables $x_1, \cdots, x_n$,
$M = (f)$ is a model for Φ i.e. $\forall x_1 \cdots \forall x_n \exists y : y[y/f(x_1, \cdots, x_n)]$ is true if, and only if $\forall x_1 \cdots \forall x_n f(x_1, \cdots, x_n)$ is true. That means $f(x_1, \cdots, x_n)$ is a tautology. Hence, there is a reduction from the *coNP*–complete tautology problem for propositional formulas. ■

4 Characterizations

For a class K of propositional formulas we want to characterize the class of *QCNF* formulas having a K–model. Therefore, we introduce a relation $\mathcal{Z}(S, K)$, where $S \subseteq$ *QCNF*. We demand that any formula in S has a K–model and that no proper subclause–subformula has such a model. That means, after removing an arbitrary occurrence of a literal in the formula the formula has no K–model. In a certain sense, the formula must be minimal. Therefore, additionally we demand that any *QCNF*, for which a K–model exists, contains a subclause–subformula in S.

Definition 3. *For a class of QCNF formulas S and a class of propositional formulas K we define: $\mathcal{Z}(S, K)$ holds if, and only if*

1. $\forall \Phi \in S$: *Φ has a K–model*
2. $\forall \Phi \in S \; \forall \Phi' \subset_{cl} \Phi : \Phi' \notin S$
3. $\forall \Phi : (\Phi$ *has a K–model* $\Longrightarrow \exists \Phi' \subseteq_{cl} \Phi : \;\; \Phi' \in S)$.

For our desired characterization we need the following definition of *minimal satisfiable* formulas.

Definition 4. *MINSAT* $:= \{\Phi \in QCNF \mid \Phi$ *is true,* $\forall\, \Phi' \subset_{cl} \Phi : \Phi'$ *is false*$\}$. *For* $\Phi \in$ *MINSAT we say* Φ *is minimal satisfiable.*

For $K \subseteq$ *CNF, a quantified Boolean formula* Φ *is in* K*–MINSAT if and only if* Φ *has a* K*–model and for any* $\Phi' \subset_{cl} \Phi : \Phi'$ *has no* K*–model.*

For example, the formula $\Phi = \forall x \forall z \exists y (z \vee x \vee y) \wedge (\neg x \vee \neg y)$ is not in *MINSAT* , because Φ is true and the subclause–subformula $\Phi' = \forall x \forall z \exists y (x \vee y) \wedge (\neg x \vee \neg y)$ is true, too. The formula Φ' is in *MINSAT* , because after removing an arbitrary literal from Φ' the formula is false.

The class *MINSAT* is often used for the relation $\Xi(S, K)$. For example, later on we will see that $\Xi(QHORN \cap MINSAT\ , HORN)$ holds.

Subsequently, we will show that in a certain sense for the relation $\Xi(S, K)$ the classes S and K are unique, if one of them is fixed. For that reason we associate every propositional formula f with a *QCNF* formula Φ_f.

Definition 5. *Let* f *be a CNF formula. A CNF formula* g *is called* minimal *for* f*, if* $g \approx f$ *and removing an arbitrary occurrence of a literal from* g *results in a formula not equivalent to* f*.*

Definition 6. *(Associated QCNF formula* Φ_f*)*
We associate to every Boolean function f *over* $x_1, \cdots, x_n$ *QCNF formulas* Φ_f *as follows:*
To the constant function 0 we associate the formula $\phi_f := \exists y\ \neg y$
to the constant 1 we associate the formula $\Phi_f := \exists y\ y$*.*
Suppose f *is not a constant. Let* $\bigwedge \alpha_i$ *be a minimal CNF formula for* $\neg f$*, and let* $\bigwedge \beta_j$ *be a minimal CNF formula for* f*.*
Then we define $\Phi_f := \forall x_1 \cdots \forall x_n\ \exists y (\bigwedge(\alpha_i \vee y) \wedge \bigwedge(\beta_j \vee \neg y))$*.*

Lemma 2. *Suppose we have* $\Xi(S, K)$ *for some* K *and* S*. Then* $\{\Phi_f | f \in K\} \subseteq S$*.*

Proof: At first we show that $M = (f)$ is a model for Φ_f.
If f is a constant then obviously M is a model for Φ_f. If f is not a constant, then:
$M = (f)$ is a model for Φ_f if, and only if
$\forall x_1 \cdots \forall x_n \exists y : \bigwedge_i (\alpha_i \vee y) \wedge \bigwedge(\beta_j \vee \neg y)[y/f(x_1, \cdots, x_n)]$ is true iff
$\forall x_1 \cdots \forall x_n \exists y : (\bigwedge_i \alpha_i \vee y) \wedge (\bigwedge \beta_j \vee \neg y)[y/f(x_1, \cdots, x_n)]$ is true iff
$\forall x_1 \cdots \forall x_n : (\bigwedge_i \alpha_i \vee f(x_1, \cdots, x_n)) \wedge (\bigwedge_j \beta_j \vee \neg f(x_1, \cdots, x_n))$ is true iff
$\forall x_1 \cdots \forall x_n : (\neg f(x_1, \cdots, x_n) \vee f(x_1, \cdots, x_n)) \wedge (f(x_1, \cdots, x_n) \vee \neg f(x_1, \cdots, x_n))$ is true.

Since the last formula is a tautology, M is a model for Φ_f.
Next we show that there is no satisfiable formula $\Phi' \subset_{cl} \Phi_f$. Suppose the

contrary, there exists such subclause–subformula Φ'. Please note, that f is not a constant.

Case 1: some y is missing.
$\Phi' = \forall x_1 \cdots \forall x_n \exists y : (\bigwedge_{i \neq i_0} (\alpha_i \vee y) \wedge \alpha_{i_0} \wedge (\bigwedge (\beta_j \vee \neg y)$
The subclause α_{i_0} is a non–tautological universal clause. Hence, the formula Φ' is not true in contradiction to our assumption.
Case 2: subclause $\alpha'_{i_0} \subset \alpha_{i_0}$ for some i_0.
$\Phi' = \forall x_1 \cdots \forall x_n \exists y (\bigwedge_{i \neq i_0} (\alpha_i \vee y) \wedge (\alpha'_{i_0} \vee y) \wedge (\bigwedge (\beta_j \vee \neg y)$
Subcase: α'_{i_0} is empty.
Then the formula has the form $\Phi' = \forall x_1 \cdots \forall x_n \exists y (\bigwedge_{i \neq i_0} (\alpha_i \vee y) \wedge (y) \wedge (\bigwedge (\beta_j \vee \neg y)$
Since the formula is true, the truth value of y must always be 1. That implies $\forall x_1 \cdots \forall x_n \bigwedge \beta_j$ is true. But then f is the constant 1 in contradiction to our assumption, because $\bigwedge \beta_j \approx f$.
Subcase: α'_{i_0} is not empty.
Then the propositional formula $(\bigwedge_{i \neq i_0} \alpha_i \wedge \alpha'_{i_0}) \vee (\bigwedge \beta_j)$ must be a tautology. That means, the complement $(\bigvee_{i \neq i_0} \neg \alpha_i \vee \neg \alpha'_{i_0}) \wedge (\neg(\bigwedge \beta_j))$ is unsatisfiable. Since $f \approx \bigwedge \beta_j$, the formula
$(\bigvee_{i \neq i_0} \neg \alpha_i \vee \neg \alpha'_{i_0}) \wedge (\neg f)$ is unsatisfiable. Hence, we have $\neg f \models (\bigwedge_{i \neq i_0} \alpha_i \wedge \alpha'_{i_0})$. From the other side $\bigwedge_{i \neq i_0} \alpha_i \wedge \alpha'_{i_0} \models \bigwedge_i \alpha_i$, because $\alpha'_{i_0} \subset_{cl} \alpha_{i_0}$. Therefore, we have $\bigwedge_{i \neq i_0} \alpha_i \wedge \alpha_{i_0} \models \neg f$. That is a contradiction to the minimality of $\bigwedge_i \alpha_i$ for $\neg f$, since α'_{i_0} is a proper subclause of α_{i_0}.
Altogether, we have shown that no proper subclause-subformula of Φ_f is true. Hence, Φ_f must be in S.

■

Definition 7. *For two classes of propositional formulas K_1 and K_2 we define: $K_1 \approx K_2$ if and only if $\forall f \in K_1 \exists g \in K_2 : f \approx g$ and vice versa.*

Lemma 3. *(Uniqueness)*
1. $\forall S_1, S_2 \subseteq QCNF \forall K \subseteq$ propositional formulas: $\mathcal{Z}(S_1, K)$ and $\mathcal{Z}(S_2, K) \Longrightarrow S_1 = S_2$
2. $\forall S \subseteq QCNF \quad \forall K_1, K_2 \subseteq$ propositional formulas: $\mathcal{Z}(S, K_1)$ and $\mathcal{Z}(S, K_2) \Longrightarrow K_1 \approx K_2$

Proof: Ad 1: Suppose $\mathcal{Z}(S_1, K)$ and $\mathcal{Z}(S_2, K)$ are given, but $S_1 \neq S_2$. Let $\Phi \in S_1 - S_2$. Then Φ has a K–model, because of $\Phi \in S_1$. Since $\mathcal{Z}(S_2, K)$, there exists a subclause–subformula $\Phi' \subseteq_{cl} \Phi$ with $\Phi' \in S_2$.
Since Φ' has a K–model, there is a formula $\Phi'' \subseteq_{cl} \Phi' \subseteq_{cl} \Phi$ with $\Phi'' \in S_1$. Because of the minimality of S_1, see definition of $\mathcal{Z}$ part 2, we obtain $\Phi = \Phi''$ and therefore $\Phi \in S_2$ in contradiction to our assumption.
Ad 2: Suppose we have $\mathcal{Z}(S, K_1)$, $\mathcal{Z}(S, K_2)$ and $K_1 \not\approx K_2$. Then there is without loss of generality a formula $f \in K_1$, for which no formula in K_2 is equivalent to f. Using lemma 2 we have $\Phi_f \in S$. Since for Φ_f any model formula g is

equivalent to f and $\mathcal{Z}(S, K_2)$, there must be a model formula $g \in K_2$ equivalent to f in contradiction to our assumption.

■

Based on the definition of the relation $\mathcal{Z}(S, K)$ now we can formulate the so called characterization problems.

Characterization Problems:

1. For a given class of propositional formulas K determine the class S with $\mathcal{Z}(S, K)$.
2. For a class $S \subseteq$ *QCNF* $\cap$ *MINSAT* determine a minimum covering $S_1, \cdots S_n$, respectively $K_1, \cdots K_n$, such that $S \subseteq \bigcup_i S_i$ and $\mathcal{Z}(S_i, K_i)$.

5 Classes of models

In this section we investigate the K–model checking and K–model problem for various classes K of Boolean functions given as propositional formulas or as quantified Boolean formulas with free variables. At first we define some classes K followed by an overview of known or later on proved complexity results.

Definition 8. *We define* ($x^0 = \neg x$ *and* $x^1 = x$)
$K_0 := \{f \mid f \text{ is } 0 \text{ or } 1\}$
$K_1 := \{f \mid \exists i \exists \epsilon \in \{0,1\} : f(x_1, ..., x_n) = x_i^\epsilon,\ n \in \mathbf{N}\} \cup K_0$
$K_2 := \{f \mid \exists I \subseteq \{1, \cdots, n\} : f(x_1, ..., x_n) = \bigwedge_{i \in I} x_i,\ n \in \mathbf{N}\} \cup K_0$

In the following tabular the first column contains classes K of Boolean functions. The last two rows are devoted two Boolean functions given as quantified Boolean formulas with free variables. The abbreviation *PSPACE*–c respectively *NP*–c and *coNP*–c stands for *PSPACE*-complete respectively *NP*–complete and *coNP*–complete. Σ_2^P is a class in the polynomial-time hierarchy

Boolean functions K	model checking	model	$\mathcal{Z}(\mathbf{S}, K)$
CNF	*coNP*-c	*PSPACE*–c	*QCNF* $\cap$ *MINSAT*
$K_.$	linear time	*NP*-c	*Q1-CNF* $\cap$ *MINSAT*
$K_.$	linear time	*NP*-c	*Q2-CNF* $\cap$ *MINSAT*
$K_.$	quadratic time	*NP*-c	*QHORN* $\cap$ *MINSAT*
1-CNF	co-*NP*-c	$\Sigma_.^P$	
1-DNF	*coNP*-c	$\Sigma_.^P$	
2-CNF	*coNP*-c	$\Sigma_.^P$	
HORN	*coNP*-c		
2-HORN	*coNP*-c	$\Sigma_.^P$	
*QCNF**	*PSPACE*-c	*PSPACE*-c	
*QHORN**	*coNP*-c		

defined as $(k \geq 0)$: $\Sigma_0^P = \Pi_0^P = P$ the class of polytime solvable problems, $\Sigma_{k+1}^P = NP^{\Sigma_k^P}, \Pi_{k+1}^P = co - \Sigma_{k+1}^P$. Thus, $NP = \Sigma_1^P$ and $coNP = \Pi_1^P$. Relationships between prefix classes of *QBF* and classes of the polynomial-time hierarchy has been shown for example in [11].

The second and the third column states the complexity of the K–model checking and the K–model problem. In the last column the characterization S with $\Xi(S, K)$ is listed.

5.1 Basic Classes $\mathbf{K_0}$ and $\mathbf{K_1}$

We start with the class K_0, that means with constant functions 0 and 1. If we substitute in a *QCNF* formula $\Phi = \forall x_1 \exists y_1 \cdots \forall x_n \exists y_n \phi$ the existential variables y_i by some $\epsilon_i \in \{0, 1\}$, then we obtain the formula $\Phi' = \forall x_1 \cdots \forall x_n \phi[y_1/\epsilon_1, \cdots, y_n/\epsilon_n]$. The formula Φ' is true if and only if $\Phi^* = \phi'[y_1/\epsilon_1, \cdots, y_n/\epsilon_n]$ is a tautology, where ϕ' is generated by removing any occurrence of literals over x_i in ϕ'. Hence, Φ has a K_0–model if and only if $\Phi_{|\exists} = \exists y_1 \cdots \exists y_n \phi_{|\exists}$ has a K_0–model.

Furthermore, a propositional formula ϕ over the variables $y_1, \cdots, y_n$ is satisfiable if and only if $\exists y_1 \cdots \exists y_n \phi \in QCNF \cap SAT$ i.e. $\exists y_1 \cdots \exists y_n \phi$ has a K_0–model.

These observations immediately imply the following propositions:

1. The K_0–model checking problem for *QCNF* is solvable in linear time, whereas the K_0–model problem for *QCNF* is *NP*–complete.
2. For every formula $\Phi \in$ *QCNF* without tautological clauses: $\Phi_{|\exists} \in SAT$ if and only if Φ has a K_0–model.
3. A formula $\Phi = Q(\alpha_1 \wedge \cdots \wedge \alpha_n) \in$ *QCNF* has a K_0–model if and only if every clause α_i contains a literal L_i over an existential variable, such that $\bigwedge_{1 \leq i \leq n} L_i$ is satisfiable.

K_1–models, that means besides the constants 0 and 1, functions of the form $f(x_1, ..., x_n) = x_i^\epsilon$ are closely related to the class *Q2-CNF*. That any satisfiable *Q2-CNF* formulas has a K_1–model follows immediately from the linear time algorithm deciding the satisfiability problem for *Q2-CNF* [8]. The other propositions of the following theorem can be shown easily. A complete proof is given in [8].

Theorem 1. *([8])*

1. *For QCNF, the K_1–model checking problem is solvable in linear time and the K_1-model problem is NP–complete.*
2. *Any formula $\Phi \in$ Q2-CNF $\cap$ SAT has a K_1–model.*
3. *A formula $\Phi \in$ QCNF has a K_1–model if, and only if there is some $\Phi' \subseteq_{cl} \Phi : \Phi' \in$ Q2-CNF $\cap$ SAT .*

5.2 QHORN Versus $\mathbf{K_2}$

The class of quantified Boolean Horn formulas *QHORN* is one of the *QBF* classes for which the satisfiability problem is solvable in polynomial time. There exist algorithms deciding the satisfiabilty problem in quadratic time [7]. In this section we present some results which show, besides the constants 0 and 1, that any satisfiable *QHORN* has monomials $f(x_1, \cdots, x_n) = \bigwedge_{i \in I} x_i$ as a model. Moreover, any *QCNF* formula having a K_2–model contains a satisfiable subclause–subformula in *QHORN*.

Theorem 2. *[8]*

1. *The K_2–model checking problem for QCNF is solvable in quadratic time.*
2. *The K_2–model problem for QCNF is NP–complete.*
3. *Any formula $\Phi \in QHORN \cap SAT$ has a K_2–model and a K_2–model can be computed in polynomial time.*
4. *A formula $\Phi \in QCNF$ has a K_2–model if and only if there is some $\Phi' \subseteq_{cl} \Phi$: $\Phi' \in QHORN \cap SAT$.*
5. $\mathcal{Z}(QHORN \cap MINSAT , K_2)$

The *intersection* of two models $M_1 = (f_1, \cdots, f_n)$ and $M_2 = (g_1, \cdots, g_n)$ is defined as $M_1 \cap M_2 = (f_1 \wedge g_1, \cdots, f_n \wedge g_n)$.

For arbitrary formulas $\Phi \in QCNF$ with two distinct models M_1 and M_2, in general the intersection $M_1 \cap M_2$ is not a model for Φ. Take for example the formula. $\Phi = \exists x \exists y (x \vee y)$ Then $(x = 1, y = 0)$ and $(x = 0, y = 1)$ are distinct models for Φ, but the intersection $M = (x = 0, y = 0)$ is not a model.

For propositional Horn formulas the intersection of models, that means satisfying truth assignments, is a satisfying truth assignment, too. A similar result holds for *QHORN*.

Theorem 3. *For $\Phi \in QHORN$, if $M_1 = (f_1, \cdots, f_r)$ and $M_2 = (g_1, \cdots, g_r)$ are models for Φ, then $M_1 \cap M_2 = (f_1 \wedge g_1, \cdots, f_r \wedge g_r)$ is a model for Φ.*

Proof: Let $\Phi = Q(\phi_1 \wedge \cdots \wedge \phi_m) \in QHORN$ be a formula with universal variables $x_1 \cdots, x_q$ and existential variables $y_1 \cdots, y_r$. For $h_i := f_i \wedge g_i (1 \leq i \leq r)$, it suffices to show that for the clauses ϕ_j $(1 \leq j \leq m)$, the propositional formula $\phi_j[y_1/h_1, \cdots, y_r/h_r]$ is a tautology. We proceed by a case distinction on the structure of Horn clauses.
Case 1: (no positive $\exists$–literal in ϕ_j)
Let $\phi_j = (w \vee \bigvee_{d \in D} \neg y_d)$, where w is a disjunction of $\forall$–literals in which at most one positive literal occurs and $D \subseteq \{1, \cdots, r\}$. We obtain
$\phi'_j := \phi_j[y_1/h_1, \cdots, y_r/h_r] = (w \vee \bigvee_{d \in D} \neg h_d) = (w \vee \bigvee_{d \in D} \neg(f_d \wedge g_d)) = (w \vee \bigvee_{d \in D} \neg f_d \vee \neg g_d)$

Since M_1 is a model for Φ, the formula $\phi_j[y_1/f_1, \cdots, y_r/f_r] = (w \vee \bigvee_{d \in D} \neg f_d)$ is a tautology. Hence, ϕ'_j is a tautology.
Case 2: (positive $\exists$–variable y_1 in ϕ_j) Let $\phi_j = (w \vee y_1 \vee \bigvee_{d \in D} \neg y_d)$, where w is

a disjunction of negative ∀–variables and
$D \subseteq \{1, \cdots, r\}$. We obtain
$\phi'_j := \phi_j[y_1/h_1, \cdots, y_r/h_r] = (w \vee h_1 \vee \bigvee_{d\in D} \neg h_d) = (w \vee (f_1 \wedge g_1) \vee \bigvee_{d\in D} \neg(f_d \wedge g_d)) =$

$(w \vee (f_1 \wedge g_1) \vee \bigvee_{d\in D}(\neg f_d \vee \neg g_d))) = (w \vee f_1 \vee \bigvee_{d\in D}(\neg f_d \vee \neg g_d))) \wedge (w \vee g_1 \vee \bigvee_{d\in D}(\neg f_d \vee \neg g_d)))$.

Since M_1 and M_2 are models for Φ, $\phi_j[y_1/f_1, \cdots, y_r/f_r] = (w \vee f_1 \vee \bigvee_{d\in D} \neg f_d)$ and $\phi_j[y_1/g_1, \cdots, y_r/g_r] = (w \vee g_1 \vee \bigvee_{d\in D} \neg g_d)$ are tautologies. Hence, ϕ' is a tautology. ∎

5.3 Classes 1-CNF, 1-DNF, 2–CNF, Horn, and 2–Horn

In this section we investigate for further classes of propositional formulas the complexity of the model problem as well as the model checking problem.

Lemma 4. *For $X \in \{$1-CNF, 1-DNF, 2-CNF, HORN, 2-HORN$\}$, the X-model checking problem is coNP–complete, whereas for $Y \in \{$1-CNF, 1-DNF, 2-CNF, 2-HORN$\}$ the Y-model problem is in Σ_2^P.*

Proof: *Model checking problem*: That the various model checking problems are in *coNP* follows from the fact that the general *CNF*–model checking is *coNP*–complete.
At first we show the *coNP*–hardness for *1-CNF*–models. The *coNP*–hardness follows from the following reduction of the tautology problem for propositional formulas in 3-*DNF*. We associate $\alpha = \bigvee_{1\leq i\leq n}(l_{i1} \wedge l_{i2} \wedge l_{i3})$, where the l_{ij} are literals over the variables $x_1, \cdots, x_m$ with the formula

$$\Phi = \forall x_1 \cdots \forall x_m \exists y_1 \cdots \exists y_n \phi \text{ with } \phi = (y_1 \vee \cdots \vee y_n).$$

For each y_i, we define $f_{y_i}(x_1, \cdots, x_m) = (l_{i1} \wedge l_{i2} \wedge l_{i3})$ and $M = (f_{y_1}, \cdots, f_{y_n})$. Clearly, $\Phi[y_1/f_{y_1}(x_1, \cdots, x_m), \cdots, y_n/f_{y_n}(x_1, \cdots, x_m)] = \forall x_1 \cdots \forall x_m \alpha$, and therefore M is a model of Φ if, and only if α is a tautology.
The *coNP*–hardness for the classes *2-CNF, HORN, 2-HORN* follows, because *1-CNF* $\subseteq$ *2-CNF, HORN, 2-HORN*.
The *coNP*–hardness for the *1-DNF*–model checking problem can be shown similarly by associating with α the formula

$$\Phi = \forall x_1 \cdots \forall x_m \exists y_1 \cdots \exists y_n \phi \text{ with } \phi = (\neg y_1 \vee \cdots \vee \neg y_n).$$

Model problem:
At first we show: there is a polynomial p, such that if a formula Φ has a *2-CNF*–model, then there is *2-CNF*–model of length less than $p(|\Phi|)$

For n variables there exists at most $4n^2 + 2n$ different 2–clauses over these variables. Therefore, the maximum *2-CNF* over n variables has length less than $2\times(4n^2+2n)$. If a *QCNF* has n universal variables, then after the substitution of t occurrences of existential variables by some *2-CNF*–formulas over these variables the length of the resulting formula is less or equal than $|\Phi|+t\times(2\times(4n^2+2n)) \leq 13|\Phi|^3$.
Since *2-HORN* and *1-CNF* is a proper subset of *2-CNF* an analogue proposition holds.

That means, we can solve the model problem for *2-CNF* respectively *2-HORN* by non–deterministically guessing a proper sequence of formulas in *2-CNF* respectively *2-HORN* of length $\leq 13|\Phi|^3$. In a second step we check whether the formulas build a model for Φ. Since the model checking problems are in *coNP* we obtain our desired result.
The proof for *1-DNF* is similar to that for *1-CNF*. Altogether, we see that the model checking problems are in Σ_2^P. ■

For *HORN* formulas the number of different clauses over n variables can not be bounded by a polynomial in n. Therefore, we can not conclude as for *2-CNF* that the *HORN*–model problem is Σ_2^P.
It is an open problem whether there is a polynomial q such that any *QCNF* formulas Φ, for which a *HORN*–model exists, has always a *HORN*–model of length less or equal than $q(|\Phi|)$.

5.4 *QCNF**–Models

In this section we discuss the case that the Boolean functions are given as quantified Boolean formulas with free variables. It is well known that a Boolean function f over the variables $x_1, \cdots, x_n$ can be represented as quantified Boolean formula with *CNF* kernel and free variables $x_1, \cdots, x_n$. For example

$$\Phi(x_{1,1}, \cdots, x_{n,n}) = \exists y_1 \cdots \exists y_n((\neg y_1 \vee \cdots \vee \neg y_n) \wedge \bigwedge_{1\leq i,j\leq n} (y_i \vee \neg x_{i,j}))$$

is logically equivalent to the propositional formula

$$\varphi = \bigwedge_{1\leq j_1,\cdots,j_n\leq n} (\neg x_{1,j_1} \vee \cdots \vee \neg x_{n,j_n})$$

The length of $\Phi(x_{1,1}, \cdots, x_{n,n})$ is $2n^2 + 2n$, whereas φ has length n^{n+1}.

In comparison to propositional formulas in some cases the length of the quantified Boolean formula with free variables is essentially shorter than the shortest *CNF* representation (see example above). The set of quantified Boolean formulas with free variables is denoted as *QCNF**. Now instead of substituting existential variables by propositional formulas we replace the existential variables by formulas in *QCNF** and adopt the K–model definitions. Please note, that a substitution of *QCNF** formulas may lead to *QBF* formula not necessary in prenex normal form. But the resulting formula contains no free variables, since the free variables of the substitute are bounded by the universal variables of the input formula.

The question whether a quantified Boolean formula has a $QCNF^*$-model is equivalent to the question whether the formula has a *CNF*-model, since any Boolean formula and therefore any *CNF* formula can be represented by a $QCNF^*$-formula. That shows the *PSPACE*-completeness of the $QCNF^*$-model problem. On the other hand, the *CNF*-model checking is *coNP*-complete (see Lemma 1), whereas the $QCNF^*$-model checking problem seems to be much harder.

Lemma 5. *The $QCNF^*$-model checking problem is PSPACE-complete.*

Proof: The problem lies in *PSPACE*, because the substitution of the existential variables y_i by quantified Boolean formulas with free variables ψ_i in a formula Φ leads to a quantified Boolean formula of lenght $\leq |\Phi|(\Sigma_i|\psi_i|)$. The problem whether the resulting *QBF* formula is true, obviously is in *PSPACE*.
Now it remains to show that the $QCNF^*$-model checking problem is *PSPACE*-hard. For the formula $\Phi = \exists y \quad y$ and an arbitrary formula $\varphi \in QCNF$: $M = (\varphi)$ is a model for Φ if and only if φ is true. Since the evaluation problem for *QCNF* is *PSPACE*-complete ([7,10]), the $QCNF^*$-model checking problem is *PSPACE*-hard. ■

Now we restrict ourselves to a subclass of $QCNF^*$ which has a satisfiability problem solvable in polynomial time. $QCNF^*$ formulas with a kernel in form of a propositional Horn formula are denoted as $QHORN^*$-formulas. The formula given in the example above is in $QHORN^*$. The following proposition states a known relationship between these formulas and propositional Horn formulas *HORN*.

Proposition 1. *(Horn equivalence)[7]*

1. $\forall \Phi \in QHORN^* \quad \exists F \in HORN : \Phi \approx F$,
2. *There exist formulas $\Phi_n \in QHORN^*$ for which any equivalent CNF formula has superpolynomial length.*

The next theorem shows that with respect to the complexity of the problems there is no big difference between Boolean functions represented as *HORN* and given as $QHORN^*$ formulas.

Theorem 4. *($QHORN^*$-models)*

1. *The problem whether a QCNF formula has a $QHORN^*$-model is as hard as the problem whether a Horn–model exists.*
2. *The $QHORN^*$-model checking problem is as hard as the HORN–model checking problem. (coNP–complete)*

Proof: Ad 1: Since any $QHORN^*$ formula with free variables $x_1, \cdots, x_n$ is equivalent to a *HORN* formula over these variables, a *QCNF* formula has a $QHORN^*$-model if, and only if the formula has a *HORN*-model.
Ad 2: We will show that the problem is *coNP*-complete.
The *coNP*-hardness of the $QHORN^*$-model checking problem follows directly

from the *coNP*–completeness of the *HORN*–model checking problem, since every *HORN* formula is a *QHORN** formula.
Now it remains to show that the problem lies in *coNP*. Let be given a *QCNF* formula
$\Phi = \forall z_1 \exists y_1 \cdots \forall z_n \exists y_n \bigwedge_{1 \leq j \leq m} \phi_j$ and a proper sequence of *QHORN** formulas $M = (\varphi_{y_1}, \cdots, \varphi_{y_n})$, where φ_{y_r} has the free variables $z_1, \cdots, z_r$.
Next we establish a poly–time non-deterministic procedure for the complementary problem, that M is not a model for Φ. We have: M is not a model for Φ if and only if

$\forall z_1 \cdots \forall z_n \bigwedge_{1 \leq j \leq m} \phi_j[y_1/\varphi_{y_1}(z_1), \cdots, y_n/\varphi_{y_n}(z_1, \cdots, z_n)]$ is false if and only if

$\exists z_1 \cdots \exists z_n \bigvee_{1 \leq j \leq m} \neg \phi_j[y_1/\varphi_{y_1}(z_1), \cdots, y_n/\varphi_{y_n}(z_1, \cdots, z_n)]$ is true.

Now we guess non–deterministically a truth assignment for the variables $z_1, \cdots, z_n$, and replace the variables z_i by these truth values. Then we obtain a *QBF* formula
$\Phi' := \bigvee_{1 \leq j \leq m} \neg \phi_j[y_1/\varphi'_{y_1}, \cdots, y_n/\varphi'_{y_n}]$, where the formulas φ'_{y_j} are the result of the replacing of the variables z_i by the truth values in $\varphi_{y_i}(z_1, \cdots, z_i)$. These formulas are closed *QHORN* formulas.
Now we show the poly–time solvability of the problem whether the formula Φ' is true.
At first we compute the truth value of all the closed *QHORN* formulas φ'_{y_j}. That can be done in poly–time, since the evaluation (satisfiability) problem for *QHORN* is solvable in poly–time. Now we replace the formulas φ'_{y_j} in Φ' by their truth values. The resulting formula is a propositional formula with the constants 0 and 1, but without variables. The evaluation of whether the formula is true costs not more than linear time.
If the result is true, then we know that Φ' is true and therefore M is not a model for Φ. Further, if M is not a model, then there is an truth assignment for z_i for which Φ' is true.
Hence, the *QHORN**–model checking is in *coNP*. ∎

6 Conclusions and Future Work

This paper opens a new approach to explore quantified Boolean formulas by studying Boolean function models. Several issues remain for further work. Actually, we do not know the complexity of the Horn–model problem and the *QHORN**–model problem. For characterizations, the classes of *QCNF* formulas, which can be characterized by *1-CNF*–models, *1-DNF*–models, *2-CNF*–models, *2-HORN*, etc. are not clear. In addition, models consisting of Boolean functions represented by quantified Boolean formulas with free variables are rather interesting, because of the space saving representations.

References

[1] Aspvall, B., Plass, M. F., and Tarjan, R. E.: A Linear-Time Algorithm for Testing the Truth of Certain Quantified Boolean Formulas, *Information Processing Letters*, **8** (1979), pp. 121-123

[2] Cadoli, M., Schaerf, M., Giovanardi, A., and Giovanardi, M.: An Algorithm to Evaluate Quantified Boolean Formulas and its Evaluation, In: *highlights of Satisfiability Research in the Year 2000*, (2000), IOS Press.

[3] Cook, S., Soltys, M.: Boolean Programs and Quantified Propositional Proof Systems, *Bulletin of the Section of Logic*, **28** (1999), pp. 119-129.

[4] Feldmann, R., Monien, B., and Schamberger, S.: A Distributed Algorithm to Evaluate Quantified Boolean Formulas, In: *proceedings of AAAI*, (2000).

[5] Flögel, A., Karpinski, M., and Kleine Büning, H.: Resolution for Quantified Boolean Formulas, *Information and Computation* **117** (1995), pp. 12-18

[6] Giunchiglia, E., Narizzano, M., and Tacchella, A.: QuBE: A System for Deciding Quantified Boolean Formulas, In: *Proceedings of IJCAR*, (2001), Siena.

[7] Kleine Büning, H., Lettmann, T.: *Propositional Logic: Deduction and Algorithms*, Cambridge University Press, (1999).

[8] Kleine Büning, H., Subramani, K., and Zhao, X.: On Boolean Models for Quantified Boolean Horn Formulas, *SAT 2003*, Italy. *Lecture Notes in Computer Science* **2919** pp. 93–104, 2004

[9] Letz R., Advances in Decision Procedure for Quantified Boolean Formulas, In: *Proceedings of IJCAR*, (2001), Siena.

[10] Meyer A.R., Stockmeyer L.J., Word Problems Requiring Exponential Time: Preliminary Report, In: *Proc.* 5th *Ann. Symp. on Theory of Computing* (1973), pp.1–9.

[11] Stockmeyer, L. J.: The Polynomial-Time Hierarchy, In: *Theoretical Computer Science*, **3** 1977, pp. 1-22.

[12] Rintanen, J.T.: Improvements to the Evaluation of Quantified Boolean Formulae, In: *Proceedings of IJCAI*, (1999).

Polynomial Algorithms for MPSP Using Parametric Linear Programming

Babak Mougouie

Max-Planck Institut für Informatik,
Stuhlsatzenhausweg 85, 66123 Saarbrücken, Germany
mbabak@mpi-sb.mpg.de

Abstract. The multiprocessor scheduling problem(MPSP), $P|prec, p_j = 1|C_{max}$, is known to be NP-complete. The problem is polynomially solvable, however, if the precedence relations are of the intree(outtree) type, $P|intree(outtree), p_j = 1|C_{max}$, or if the number of processors is two, $P2|prec, p_j = 1|C_{max}$. In this paper, we introduce a parametric linear program which gives a lower bound for the makespan of MPSP and retrieves the makespans of the two polynomially solvable problems.

1 Introduction

Let $G = (V, E)$ be an acyclic directed graph. We denote by $V = \{v_1, \ldots, v_n\}$ the set of tasks, and by E the precedences between those tasks. A precedence $e = \langle v_i, v_j \rangle$ implies that the task v_j depends on the result of the task v_i and v_i must be completed before v_j. If there is a path $\langle v_i, ..., v_j \rangle$ from v_i to v_j in G, we say v_i is a *predecessor* of v_j and v_j is a *successor* of v_i. Let $Pred(v_i)$ and $Succ(v_i)$ denote the set of all predecessors and successors of v_i consecutively.

Furthermore let $P = \{p_1, \ldots, p_\sigma\}$ be a set of identical processors. Each task v_i can be processed by one of the processors $p \in P$ in 1 unit time; therefore we can define a discrete set of time-steps $\Gamma = \{1, \ldots, n\}$. No two tasks can be assigned to the same processor at a time-step t. A *schedule* $\mathcal{S} = (x_{1,1}, ..., x_{n,n})$ is a mapping of V to Γ. We denote $x_{i,t} = 1$ when a task v_i is scheduled to a processor at the time-step $t \in \Gamma$ and $x_{i,t} = 0$ otherwise.

The *assignment constraints* are defined as $\sum_{t \in \Gamma} x_{i,t} = 1 \quad \forall v_i \in V$ which means each task should be processed once. The *resource* and *precedence constraints* are defined as $\sum_{v_i \in V} x_{i,t} \leq \sigma \quad \forall t \in \Gamma$ and $\sum_{t \geq \gamma} x_{i,t} + \sum_{t \leq \gamma} x_{j,t} \leq 1 \quad \forall \langle v_i, v_j \rangle \in E$ and $\forall \gamma \in \Gamma$.

Additionally, we define a positive integer variable T such that the *time constraints* $x_{i,t}.t \leq T \quad \forall v_i \in V$ and $\forall t \in \Gamma$ are satisfied. The multiprocessor scheduling problem **(MPSP)** seeks to minimize T such that the assignment, resource, time and precedence constraints are satisfied.

W. Lenski (Ed.): Logic versus Approximation, LNCS 3075, pp. 33–42, 2004.

MPSP is formulated as:

$$
\begin{array}{llll}
\textbf{min} & T & & \\
 & & & \\
\text{s.t.} & \sum_{t\in\Gamma} x_{i,t} = 1 & \forall v_i \in V & \\
 & \sum_{v_i\in V} x_{i,t} \le \sigma & \forall t \in \Gamma & \\
 & x_{i,t}.t \le T & \forall v_i \in V & \& \ \forall t \in \Gamma \\
 \sum_{t\ge\gamma} x_{i,t} + & \sum_{t\le\gamma} x_{j,t} \le 1 & \forall \langle v_i, v_j\rangle \in E & \& \ \forall \gamma \in \Gamma \\
 & x_{i,t} \in \{0,1\} & \forall v_i \in V & \& \ \forall t \in \Gamma \\
 & T \in I\!N. & &
\end{array}
\tag{1}
$$

T^*(to be clearer sometimes we write $T^*(G)$), the optimal objective value of (1), is the *makespan* of MPSP which is the last time-step that a node in G is scheduled. A schedule $\mathcal{S}$ is *valid* if it satisfies (1) and S^*, the optimal solution corresponding to T^*, is called the *optimal schedule.* Obviously, an optimal schedule must not be unique.

MPSP is known to be NP-complete in general [9]. However, a number of special cases can be solved in polynomial time. In [7], Hu presented his "level algorithm" which is applied to solve MPSP when G is either an *intree*, i.e., each node in G has at most one immediate successor, or an *outtree*, i.e., each node in G has at most one immediate predecessor(if $\langle v_i, v_j\rangle \in E$, then v_j is the *immediate successor* of v_i and v_i is the *immediate predecessor* of v_j).

Fuji, Kasami & Ninomiya [3] presented the first polynomial algorithm for MPSP when $\sigma = 2$ based on matching techniques. Coffman & Graham [1] gave another algorithm based on a lexicographic numbering scheme. The run-time of their algorithm was improved by Sethi [8], Gabow [4] and Gabow & Tarjan [5].

In this paper, we present a parametric linear program, which gives a lower bound for the makespan of MPSP and additionally provides the makespans of the cases G = intree(or outtree) and $\sigma = 2$. This unifies the two methods, Hu's level algorithm, HLA [7], and Highest level first algorithm, HLF [4].

The paper is organized as follows: In section 2 we introduce the parametric linear program, PLP, and show that it's optimal objective value is a lower bound for the makespan of MPSP. In section 3 we present a modified version of HLA, and prove that applying PLP and HLA on intrees(or outtrees) we get the same makespan; therefore the bound found by PLP is tight. In section 4 we find the makespan of $P2|prec, p_j = 1|C_{max}$ using a modification of HLF and PLP. In section 5, the generalitry of PLP is outlined. Section 6 briefly summarizes the results and gives an outlook to further research.

2 The Parametric Linear Program(PLP)

Definition 1. *PLP is defined as:*

$$\mathit{min}\ T \quad \mathit{s.t.}\ \mathcal{P}(\lambda) = (LP(1) \cap \{x_{i,t} = 0 \quad \forall v_i \in V, \forall t > \lambda\})$$

where $LP(1)$ is the linear relaxation of (1) and λ is an integer parameter. Let $X = (x_{1,1}, ..., x_{n,n}) \in \mathcal{P}(\lambda)$ and λ^* be the smallest integer such that $\mathcal{P}(\lambda^*)$ is not empty. To be clearer, we write sometimes $\lambda^*(G) = \lambda^*$.

Definition 2. *The following terminologies are used throughout this paper.*

- CP: critical path *of* G = *longest path in* G; *and* $CP(G) = length(CP)$.
- $length(\rightarrow v_i)$: *length of the longest path from any node in* G *to* $v_i \in G$.
- $length(v_i \rightarrow)$: *length of the longest path from* $v_i \in G$ *to any node in* G.
- $\overline{\alpha}(v_i) = length(\rightarrow v_i) + 1; \quad \underline{\alpha}(v_i) = length(v_i \rightarrow)$;
- $\alpha = \alpha(G) = CP(G) + 1$.

- $U_k = \{v_i | v_i \in V \quad \& \quad \overline{\alpha}(v_i) = k\} \quad \forall k = 1, ..., \alpha.$
  $L_k = \{v_i | v_i \in V \quad \& \; \alpha - \underline{\alpha}(v_i) = k\} \quad \forall k = 1, ..., \alpha.$
  U_k and L_k were introduced for the first time by Fernandez & Bussel [2]. It's obvious that U_k (and L_k) are α nonempty disjoint partitions of V where
  $$V = (\cup_{k=1}^{\alpha} U_k) = (\cup_{k=1}^{\alpha} L_k).$$
  Each U_k (or L_k) is called the successive level of U_l (respectively L_l) if $l < k$. U_k (or L_k) is called the direct successive level of U_{k-1} (or L_{k-1}) for $k = 2, ..., \alpha$.
- Decomposition *of $G = (V, E)$: Let $V = V_1 \cup V_2$. G can be* decomposed *into two graphs $G_1 = (V_1, E_1)$ and $G_2 = (V_2, E_2)$ such that $E_1 = E \backslash \{e | e \in E$ has both end points in $V_1\}$ and $E_2 = E \backslash \{e | e \in E$ has both end points in $V_2\}$.*

We start with some lemmas to prove that λ^* is a lower bound for T^*.

Lemma 1. *Let $\mathcal{P}(\lambda_1) \neq \emptyset$. Then for any $X = (x_{1,1}, ..., x_{n,n}) \in \mathcal{P}(\lambda_1)$:*

$$x_{i,u} + ... + x_{i,l} = 1 \quad \forall v_i \in V,$$

where $u = \overline{\alpha}(v_i)$ and $l = \lambda_1 - \underline{\alpha}(v_i)$.
In other words, $x_{i,t} = 0 \quad \forall t = 1, ..., u-1 \quad \& \quad t = l+1, ..., \lambda_1$.

Proof. We do the proof by induction. Let $u = \overline{\alpha}(v_i) = 2$, therefore $length(\rightarrow v_i) = 1$. Let $\langle v_j, v_i \rangle \in E$ be this longest path.

$$\left.\begin{array}{ll} \text{Precedence constraints} \Rightarrow & x_{j,1} + ... + x_{j,n} + x_{i,1} \leq 1 \\ \text{Assignment constraints} \Rightarrow & x_{j,1} + ... + x_{j,n} \qquad\quad = 1 \end{array}\right\} \Rightarrow x_{i,1} = 0.$$

Let $u = \overline{\alpha}(v_i)$ and suppose the hypothesis holds for all v_j with $\overline{\alpha}(v_j) < u$. Let $\langle ..., v_j, v_i \rangle$ be the longest path from any node in G to v_i. It's obvious that $\overline{\alpha}(v_j) = u - 1$.

$$\left.\begin{array}{lll} \text{Precedence constraints} \Rightarrow & x_{j,u-1} + ... + x_{j,n} + x_{i,1} + ... + x_{i,u-1} & \leq 1 \\ \text{Assignment constraints} \Rightarrow & x_{j,1} + ... + x_{j,n} & = 1 \\ \text{Hypothesis} \Rightarrow & x_{j,t} = 0 \qquad\qquad \forall t = 1, ..., u-2 & \end{array}\right\} \Rightarrow$$

$x_{i,t} = 0 \quad \forall t = 1, ..., u-1 \Rightarrow x_{i,u} + ... + x_{i,n} = 1.$

With the same argumetation and the fact that $x_{i,t} = 0 \quad \forall t > \lambda_1$, it can be proven $x_{i,t} = 0 \quad t = l+1, ..., \lambda_1$ where $l = \lambda_1 - \underline{\alpha}(v_i)$. □

Lemma 2. *$max(\alpha, \lceil |V|/\sigma \rceil) \leq \lambda^* \leq T^*$.*

Proof. Let $X = (x_{1,1}, ..., x_{n,n}) \in \mathcal{P}(\lambda^*)$.

- Suppose $\mathcal{P}(\alpha - 1) \neq \emptyset$. Then $\forall X \in \mathcal{P}(\alpha-1), x_{i,t} = 0 \quad \forall v_i \in V, \forall t \geq \alpha$. Let $CP = \langle ..., v_i \rangle$, therefore $\overline{\alpha}(v_i) = \alpha$. From lemma 1 we know $x_{i,t} = 0 \quad \forall t < \alpha$ and therefore $x_{i,t} = 0 \quad \forall t \in \Gamma$, which is a contradiction to assignment constraints. Thus, $\mathcal{P}(\alpha - 1) = \emptyset$ and $\alpha \leq \lambda^*$.
- Let $S^* = (x^*_{1,1}, ..., x^*_{n,n})$ be an optimal schedule. Obviously $S^* \in LP(1)$. $x^*_{i,t} = 0 \quad \forall v_i \in V, \forall t > T^* \Rightarrow S^* \in \mathcal{P}(T^*)$. Since λ^* is the smallest integer s.t. $\mathcal{P}(\lambda^*)$ is not empty, therefore $\lambda^* \leq T^*$.
- From resource constraints we have $\sum_{v_i \in V} x_{i,t} \leq \sigma \quad \forall t \in \Gamma$. $x_{i,t} = 0 \forall t > \lambda^*$, therefore $\sum_{t \in \Gamma} \sum_{v_i \in V} x_{i,t} \leq \sigma\lambda^*$. This alongside with assignment constraints will give $|V| \leq \sigma\lambda^*$. Since λ^* is integer, we get $\lceil |V|/\sigma \rceil \leq \lambda^*$.

□

Theorem 1. *$max(\lceil (\sum_{k=\gamma}^{\alpha} |U_k|)/\sigma \rceil + (\gamma - 1),\ \lceil (\sum_{k=1}^{\gamma} |L_k|)/\sigma \rceil + (\alpha - \gamma)) \leq \lambda^*$ for $\gamma = 1, ..., \alpha$.*

Proof. Let $V_1 = \cup_{k=\gamma}^{\alpha} U_k$, $V_2 = V \backslash V_1$ and $X = (x_{1,1}, ..., x_{n,n}) \in \mathcal{P}(\lambda^*)$. From the definition of U_k we have $\overline{\alpha}(v_i) \geq \gamma \quad \forall v_i \in V_1$. This and lemma 1 result $x_{i,t} = 0 \quad \forall v_i \in V_1, \forall t < \gamma$.

Let's decompose G into $G_1 = (V_1, E_1)$ and $G_2 = (V_2, E_2)$. Lemma 2 states that $\lambda^*(G_2) \geq \gamma - 1$, therefore

$$\lambda^* = \lambda^*(G) \geq \gamma - 1 + \lambda^*(G_1)$$

because $x_{i,t} = 0 \quad \forall v_i \in V_1, \forall t < \gamma$. Since $\lambda^*(G_1) \geq \lceil |V_1|/\sigma \rceil$ we get the result. The other case can be proven similarly. □

3 Hu's Level Algorithm(HLA)

Up to now, we have found a lower bound for λ^* and T^*. In this section, we investigate the tightness of this bound. To do so, we modify HLA [7] on intrees and show that the optimal schedule obtained with this algorithm has makespan $\max(\lceil (\sum_{k=\gamma}^{\alpha} |U_k|)/\sigma \rceil + (\gamma - 1),\ \lceil (\sum_{k=1}^{\gamma} |L_k|)/\sigma \rceil + (\alpha - \gamma)) = \lambda^* = T^*$ for $\gamma = 1, ..., \alpha$.

Let $G = (V, E)$ be an intree and L_k for $k = 1, ..., \alpha$ are given. A *starting node* is a node with no predecessor in G.

The algorithm "executes" $L_1, ..., L_\alpha$ starting from the smallest indeces. More precisely, suppose $L_1, ..., L_{k-1}$ have already been executed. L_k is executed as follows. σ arbitrary nodes from L_k are scheduled at the current time-step(1 in the beginning) and removed from L_k and G; and the time-step is incremented once. The same procedure is done $0 \leq |L_k| < \sigma$. If $|L_k| = 0$ the execution of L_k is completed. Otherwise the algorithm schedules starting nodes from the successive level(s) of L_k with maximum $\underline{\alpha}(v)$ until σ nodes are scheduled at the current time-step or no more starting node exists.

Input: an intree $G = (V, E)$;
$t = 1; k = 1;$
While $k \leq \alpha$ **do**
{
  X_t= arbitrary σ elements of L_k;
  E_t = edges adjacent to the nodes in X_t;
  $G = (V \backslash X_t, E \backslash E_t)$;
  $L_k = L_k \backslash X_t$;
  If $(|X_t| < \sigma)$ **then**
  {
    $X_t = X_t \cup \{\sigma - |X_t|$ starting nodes, v, with maximum $\underline{\alpha}(v)\}$;
    Remove all of these starting nodes, v, from L_l and G where $v \in L_l$;
  }
  If $(|L_k| == 0)$ **then** $k++$;
  $t++$;
};
$T' = t - 1$;
Output: X_t = the set of nodes scheduled at time-step $t = 1, \ldots, T'$.

Figure 1 gives an intree and a schedule as Gantt chart(the tth column shows the nodes in X_t).

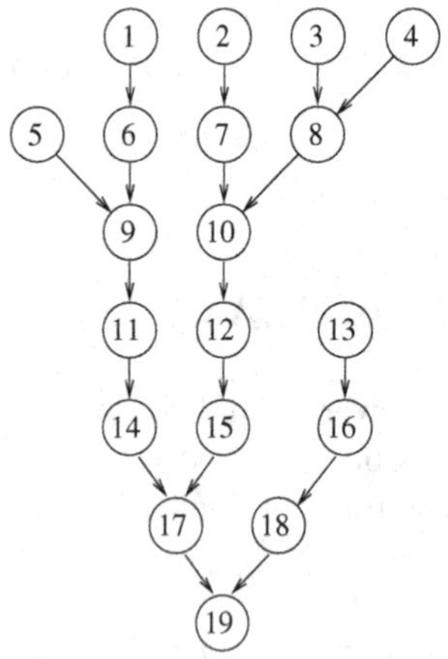

2	1	6	9	11	14	17	19
3	5	8	10	12	15		
4	7	13	16	18			

Fig. 1.

Using HLA, a schedule $X = (x_{1,1}, \ldots, x_{n,n})$ where $x_{i,t} = 1$ if $v_i \in X_t$ and $x_{i,t} = 0$ otherwise, and T' are obtained. Now we prove $\lceil(\sum_{k=\gamma}^{\alpha} |L_k|)/\sigma\rceil + (\alpha - \gamma) = T'$, which implies $\lambda^* = T' = T^*$; therefore, first, HLA gives an optimal schedule for MPSP on intrees and second, λ^* is tight. The next lemma is needed to prove this.

Lemma 3. *In the solution retrieved by HLA on G=intree, $|X_{t+1}| \leq |X_t| \leq \sigma$ for all $t = 1, \ldots, T' - 1$. Furthermore, if $|X_t| < \sigma$ for any $t = 1, \ldots, T'$, then $\exists v \in X_t$ s.t. $t = T' - \underline{\alpha}(v)$.*

This means for any $v' \in X_{\tau+1} \cup \ldots \cup X_{T'}$, there exists $v \in X_\tau$ s.t. $v \in Pred(v')$.

Proof. Obviously $|X_t| \leq \sigma \forall t = 1, \ldots, T'$. Let τ be the smallest time-step for which $|X_\tau| < \sigma$. Since G is an intree, therefore the set of immediate successors of the tasks in X_τ has at most $|X_\tau|$ members and consequently, $X_{\tau+1}$ has at most $|X_\tau|$ members. Similarly, we can prove $|X_{t+1}| \leq |X_t| \leq \sigma$ for $t = 1, \ldots, T' - 1$.

As a result, $|X_t| < \sigma$ for $t = \tau, ..., T'$. This means for any $w' \in X_{t+1}$ there exists $w \in X_t$ s.t. $\langle w, w'\rangle \in E$ for $t = \tau, \ldots, T' - 1$ which means there exists a path $\langle v_t, \ldots, v_{T'}\rangle$ with length $T' - t$ where $v_t \in X_t$ for $t = \tau, \ldots, T'$. Therefore $\exists v_t \in X_t$ s.t. $\underline{\alpha}(v_t) \geq T' - t$ for $t = \tau, ..., T'$. Additionally, $\underline{\alpha}(v_t) > T' - t$ is a contradiction because in this case, v_t cann't be scheduled at time-step t. Thus $\exists v_t \in X_t$ for all $t = \tau, \ldots, T'$ s.t. $t = T' - \underline{\alpha}(v_t)$. $|X_t| = \sigma$ for $t < \tau$, so the proof is complete. □

Theorem 2. $G = intree$: $\lceil(\sum_{k=1}^{\gamma} |L_k|)/\sigma\rceil + (\alpha - \gamma) = \lambda^* = T^*$,
$G = outtree$: $\lceil(\sum_{k=\gamma}^{\alpha} |U_k|)/\sigma\rceil + (\gamma - 1) = \lambda^* = T^*$
for $\gamma = 1, ..., \alpha$.

Proof. Let X_t for $t = 1, ..., T'$ be the output of HLA algorithm and τ be the smallest time-step that $\exists v \in X_\tau$ s.t. $\tau = T' - \underline{\alpha}(v)$. Let $\underline{\alpha}(v) = \alpha - \gamma$, therefore $T' = \tau + \alpha - \gamma$. Obviously:

- $\underline{\alpha}(v_i) < \underline{\alpha}(v) \Rightarrow \alpha - \underline{\alpha}(v_i) > \alpha - \underline{\alpha}(v) = \gamma \quad \forall v_i \in X_t \& t > \tau$.
- $\underline{\alpha}(v_i) \geq \underline{\alpha}(v) \Rightarrow \alpha - \underline{\alpha}(v_i) \leq \alpha - \underline{\alpha}(v) = \gamma \quad \forall v_i \in X_t \& t < \tau$.

As a consequence, $\cup_{k=1}^{\gamma} L_k = \{v_i | v_i \in X_t \forall t < \tau\} \cup \{v_i | v_i \in X_\tau \& \underline{\alpha}(v_i) = \underline{\alpha}(v)\}$. $|\{v_i | v_i \in X_t \forall t < \tau\}| = (\tau - 1) \times \sigma$ otherwise $\exists X_t$ s.t. $|X_t| < \sigma$ for $t < \tau$ and according to lemma 3, $\exists v \in X_t$ s.t. $t = T' - \underline{\alpha}(v)$ which is a contradiction to the fact that τ is the smallest time-step with this property.

Furtheremore, let $|\{v_i | v_i \in X_\tau \& \underline{\alpha}(v_i) = \underline{\alpha}(v)\}| = m$, where $m \leq \sigma$. This means $| \cup_{k=1}^{\gamma} L_k| = (\tau - 1) \times \sigma + m \Rightarrow | \cup_{k=1}^{\gamma} L_k|/\sigma = (\tau - 1) + m/\sigma \Rightarrow \lceil| \cup_{k=1}^{\gamma} L_k|\rceil/\sigma = \tau$. Since $T' = \tau + \alpha - \gamma$, the proof is complete.

The case of outtrees is proven similarly. □

As a result, the time complexity to find the makespan of $P|intree(outtree), p_j = 1|C_{max}$ is $O(PLP)$.

4 Level Algorithm for $\sigma = 2$(LA2)

In this section, we present LA2 algorithm which is a modification of HLF. The constraints resulted from LA2 will be added to PLP and this gives the makespan of $P2|prec, p_j = 1|C_{max}$.

First we mention the Higest level algorithm, HLF [4]. Similar to HLA, HLF "executes" $L_1, ..., L_\alpha$ starting from the smallest indeces. Let $L_1, ..., L_{k-1}$ be already executed, and L_k contains d unexecuted nodes. L_k is executed in the next $\lceil d/2 \rceil$ time-steps, as follows. Let t_0 be the current time-step, then at each time-step $t_0, ...t_0 + \lfloor d/2 \rfloor - 1$ two nodes of L_k are scheduled and then removed from L_k and G. If d is even, this completes the execution of L_k. Otherwise(d is odd) the last node of L_k is scheduled at $t_0 + \lceil d/2 \rceil$, and possibly(but not necessarily) a starting node v from the successive level(s) of L_k (we say v jumps). This completes the execution of L_k.

The last node in L_k is chosen such that the starting node v which jumps belongs to a successive level of L_k with the smallest index(or higest level as named in [4]). This guarantees the optimality of the schedule. The difficulty in constructing such a schedule arises when there is a choice of starting nodes in a successive level to jump. Our goal is to retrieve the makespan of the schedule but not the schedule itself, therefore in LA2, we just keep the track of those successive levels containing a candidate starting node to jump and we don't schedule or jump any node.

The main task of LA2 is to find in which time-steps of an optimal schedule, only one node is scheduled. Then some constraints $(*)$ are added to $\mathcal{P}(\lambda)$ to maintain this for any integral solution in $\mathcal{P}(\lambda)$. This increases λ^* that much that $\lambda^* = T^*$. LA2 works as follows:

Let $L_1, ..., L_{k-1}$ be executed and t_0 be the current time-step. Remove all nodes in L_k and their adjacent edges from G. If $|L_k| - jump_k$ is even($jump_k$ is the number of nodes already needed to jump from L_k), no action is done. Otherwise, find a starting node $v \in L_l$ $(l > k)$ with maximum $\underline{\alpha}(v)$ such that $\exists v_k \in L_k$ s.t. $v \notin Succ(v_k)$. Then we might need to do three actions:

1. If $L_l(jump)$ is empty, this means it is the first time that a node in L_l needs to jump; therefore all candidate nodes to jump in L_l are added to $L_l(jump)$.
2. If $|L_l(jump)| - jump_l = 1$, then there exists only one node to jump; therefore we remove all nodes in $L_l(jump)$ from L_l and G. Otherwise, $jump_l$ is incremented once.
3. If there exists no starting node to jump, then no node in $L_{k+1} \cup ... \cup L_\alpha$ can be scheduled at the current time-step and before that. Then some constraint $(*)$ is retrieved and added to $\mathcal{P}(\lambda)$.

The algorithm terminates when L_α is executed.

Input: $G = (V, E)$;
$t_0 = 0; D = 0;$
$L_k(jump) = \emptyset, jump_k = 0 \quad \forall k = 1, ..., \alpha;$
For $k = 1, \ldots, \alpha$ **do**

```
{
  E_k = edges adjacent to nodes in L_k;
  G = (V\L_k, E\E_k);
  D = D + |L_k| - jump_k;
  If (|L_k| - jump_k is odd) then
  {
    Find a starting node v ∈ L_l:
      - α(v) is maximum,
      - ∃v_k ∈ L_k s.t. v ∉ Succ(v_k);
    If (L_l(jump) == ∅) then
    { L_l(jump)= the set of all candidate starting nodes in L_l to jump };
    If (|L_l(jump)| - jump_l == 1) then
    { Remove L_l(jump) from G and L_l}
    else jump_l ++;
    If (v does not exists) then
    {
      x_{i,t} = 0   ∀v_i ∈ L_{k+1} ∪ ... ∪ L_α &t = 1, ..., t_0 + ⌈D/2⌉;        (*)
      t_0 = t_0 + ⌈D/2⌉;
      D = 0;
    }
    else D ++;
  }
};
```

Output: Constraints $(*)$.

Theorem 3. *Let λ^* be the smallest integer such that $\mathcal{P}(\lambda^*) \cap (*) \neq \emptyset$. Then $\lambda^* = T^*$ where T^* is the makespan of MPSP with $\sigma = 2$.*

Proof. Let γ be the first index such that $|L_\gamma| - jump_\gamma$ is odd and the next startingt node v does not exist. Let $V_1 = \cup_{k=1}^{\gamma} L_k \cup \{\text{jumped nodes } v | v \in \cup_{k>\gamma} L_k\}$. It can be easily seen from the course of the algorithm that $D = |V_1|$ when the constraint $(*)$ is retrieved for the first time. We know, as long as a starting node to jump exists, there exists an optimal schedule that in each time-step $t = 1, ..., \lfloor D/2 \rfloor$ schedules exactly two nodes(HLF guarantees that such an optimal schedule exists). Additionally, in time-step $\lceil D/2 \rceil$ of this optimal schedule is only one node scheduled. Therefore, for a subgraph $G_1 = (V_1, E_1)$ where $E_1 = \{\text{edges with both endpoints in } V_1\}$, we have $\lambda^*(G_1) = \lceil |V_1|/2 \rceil$.

Consider the graph $G_2 = (V_2, E_2)$ where $V_2 = L_{\gamma+1}, ..., L_\alpha$ and $E_2 = \{\text{edges with both endpoints in } V_2\}$. Since no starting node v exists, the jumped nodes are all removed from V_2 and therefore $V_2 = V \backslash V_1$. HLF is an optimal schedule and no node from $L_{\gamma+1}, ..., L_\alpha$ can be scheduled at time-step $\lceil D/2 \rceil$ and before that, therefore by adding the constraint $x_{i,t} = 0 \quad \forall v_i \in V_2 \& t = 1, ..., \lceil D/2 \rceil$ to $\mathcal{P}(\lambda)$, we get $\lambda^* \geq \lceil D/2 \rceil + \lambda^*(G_2)$.

Doing the same procedure we will get a sequence, $D_1 = D, ..., D_r$, such that $\lambda^* \geq \lceil D_1/2 \rceil + ... + \lceil D_r/2 \rceil$. Since HLF gives an optimal schedule, therefore $\lceil D_1/2 \rceil + ... + \lceil D_r/2 \rceil = T^*$ and because $\lambda^* \leq T^*$ we get $\lambda^* = T^*$. □

We turn our attention to the efficiency of the algorithm. The construction of L_k for $k = 1, ..., \alpha$ is done in $O(n)$. The overall time for the algorithm is dominated by the time spent in finding a starting node which is done in the worst case in $O(|E| + n)$.

5 Remarks on the Generality of PLP

The simplest natural generalization of the graphs solvable by HLA and PLP is *opposing forest.* A graph $G = (V, E)$ is an opposing forest if it can be decomposed into two disjoint and independent subgraphs, $G_I = (V_I, E_I)$ and $G_O = (V_O, E_O)$, where G_I in an intree and G_O is an outtree.

$P|opposingforest, p_j = 1|c_{max}$ is solved in polynomial time if σ is fixed [6]. This time is bounded by $O((\log n)n^{\sigma^2+2\sigma-5})$ which is clearly not efficient.

The first question is whether $\lambda^* = T^*$ or not. For the following opposing forest with $\sigma = 9$, we have $\lambda^* = 8$ but $T^* = 9$,

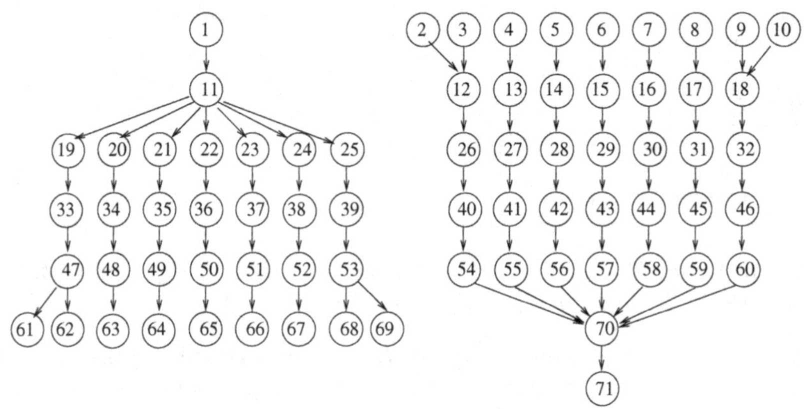

Fig. 2.

and an optimal schedule would be:

1	2		48	62
3	11		49	63
4	12		50	64
5	13		51	65
6	14	18,...,47,54,...,61	52	66
7	15		53	67
8	16		61	68
9	17		70	69
10				71

where the first two columns contain the nodes scheduled at time-steps 1 and 2 and the last two columns contain the nodes scheduled at time-steps 8 and 9. All the nodes in between can be scheduled in time-steps 3,...,7.

The problem is that all nodes in $U_2 = \{11, ..., 18\}$ can be scheduled in time-step 2, but for a feasible schedule at least one of them should be at a time-step after 2. The same problem is for L_6.

6 Summary

In this paper, we have presented a parametric linear program, PLP, to retrieve a lower bound for the makespan of multiprocessor scheduling problem(MPSP), $P|prec, p_j = 1|C_{max}$ which is NP-complete in general.

However, for the special case $P|intree(outtree), p_j = 1|C_{max}$, there exists a polynomial algorithm HLA. We have proved that the makespan found by HLA is equal to the optimal solution of PLP. Therefore, PLP provides a tight lower bound for the makespan of MPSP.

Besides, there exist several polynomial algorithms for the case $P2|prec, p_j = 1|C_{max}$ offered in [3], [1], [8], [4] and [5]. We have modified HLF algorithm presented by Gabow [4] , and the constraints obtained from our modified algorithm, LA2, will be added to PLP, which gives the makespan of MPSP with $\sigma = 2$.

Finally, the generality of PLP is considered and shown that for the case $P|opposingforest, p_j = 1|C_{max}$, PLP will not give the makespan. This will be an important part of our future work.

References

1. Coffman, E.G. Jr.; Graham, R.L.; Optimal Scheduling for Two-Processor Systems, *Acta Informatica* 1, 200-213, 1972.
2. Fernandez, E.B.; Bussel, B.; Bounds on the Number of Processors and Time for Multiprocessor Optimal Schedules, *IEEE Trans. on Comput.*, 22, 745-751, 1973.
3. Fuji, M.; Kasami, T.; Ninomiya, K.; Optimal sequencing on two equivalent processors, *SIAM J. Appl. Math.*, 17, 784-789, 1969. *Erratum*, 20, p. 141, 1971.
4. Gabow, H.N.; An Almost-Linear Algorithm for Two-Processor Scheduling, *J. Assoc. Comput. Mach.*, 29, 766-780, 1982.
5. Gabow, H.N.; Tarjan, R.E.; A Linear-Time algorithm for a Special Case of Disjoint Set Union, *J. Comput. System Sci.*, 30, 209-221, 1985.
6. Garey, M.R.; Johnson, D.S; Tarjan, R.E.; Yannakakis, M.; Scheduling opposing forests, *SIAM J. Alg. Disc. Math.* 4, 72-93, 1983.
7. Hu, T.C.; Parallel Sequencing and Assembly Line Problems, *Operation Research* 9, 841-848, 1961.
8. Sethi, R.; Scheduling Graphs on Two-Processors, *SIAM J. Compute.*, 5, 73-82, 1976.
9. Ulmann, J.D.; NP-Complete Scheduling Problems, *J. Comput. System Sci.*, 10, 384-393, 1975.

Discrete and Continuous Methods of Demography*

Walter Oberschelp

Rheinisch-Westfälische Technische Hochschule Aachen

Abstract. A discrete model of population growth in the spirit of Fibonacci's rabbits, but with arbitrary fixed times for the beginning and the end of fertility and for death is investigated. Working with generating functions for linear recursions we pursue the idea to give asymptotic estimates for the number of existing individuals by means of the powers of one single *main root* z of the function's denominator. The mathematical problem of mortality is easy to handle. While the outlines of such a paradigm are recognizable in the case of perpetual fertility, there remain open problems with the localization of roots on unexpected "bubbles", if fertility gets lost at a finite time. Therefore, an alternative method of asymptotic approximation via convolution techniques is given.
A generalization of this model to realistic situations with age dependent fertility rates is straightforward. Modern computing techniques admit a convenient survey over the existing roots. In competition with continuous models of demography the results seem to clarify the global influence of the demographic data in the so called *stable* models of demography. This model is basic for prognostics, when – more general – dynamic *changes* of the demographic parameters occur.

1 Introduction

The German and moreover the European society is threatened by a demographic danger: Senescence is increasing, childrens are rare, pensions will overtax the working young generation very soon. The predictions for the next 50 years are gloomy. The reasons for this development are obvious: Too many women give birth to only few children or none at all. The question is: In which way influence the demographic parameters, which are measured by our statistic boards, this evolution? Can we estimate the future development from these data in a transparent and reliable way?

The mathematical problem behind this seems trivial in principle: We only have to collect the birth-, mortality- and migration-statistics in an obvious way; then in pursuing each cohort separately and summing up we get the desired data; and in changing the assumptions, e.g. by considering alternative family-policies, this obvious model always admits prognostics. Thus we are in a much

* Dedicated to Michael M. Richter in remembering our joint activities to develop informatics at Aachen on a mathematical basis

W. Lenski (Ed.): Logic versus Approximation, LNCS 3075, pp. 43–58, 2004.

better position than T.R. Malthus, who in his famous "Essay on the principle of population" (1798) was confronted with the opposite effect of *over*-population. Malthus recommended decreasing the standard of living for the poor and keeping them under control in workhouses as the best remedy.

Unfortunately the program of population prognostics by elementary processing of statistical parameters lacks *global* understanding. As a typical method of experimental mathematics it calculates the reason-result-relation by brute force, but the relation between input parameters and output data is *not transparent.* Therefore one is looking for models, which give a better evidence. Obviously, only the generative behavior of *women*, of mothers and their daughters, is significant, while male persons are not essential for the I-O-analysis, even though they are not completely dispensable; but from a statistical point of view male men are merely catalysators. This remark is important for societies like China, where – as we know – much more female than male embryons are aborted.

A realistic model has to be rather complex: Surely, the *average* number of daughter-births for a woman will *not* be a reliable basis, for obviously if all women would decide to get their daughter with 40 years (and not with 20) this would be of strong influence.

The data in our context are, of course, *discrete*, since we usually have collective annual statistics with respect to some key-date. And pregnancy is a 0-1-event! Nevertheless, following the well-approved techniques of natural sciences one would try to develop a *continuous* model. In fact, this has been done in the first half of the last century in renewal theory. (An excellent source of information about this development is given in [12].) Starting with A.J. Lotka, mathematicians like W. Feller (1941) have given such an approach ([12], p. 133), and usual textbooks on population dynamics (cf. [7], (chap. 3)) seem to favor the continuous concepts. L.C. Cole (1954) has analysed the connections between both methods ([12], p. 190). The continuous existence theorems in terms of certain definite integrals are elegant, but do not give satisfactory intuitions. Absurd statements like "there are 0.006 children born worldwide in a millisecond" cannot arise in a discrete model. On the other hand there is a general experience, that things become seriously difficult in discrete models. But nevertheless, motivated by famous historic pioneers, we shall develop a discrete approach, which is able to investigate (under admittedly idealized assumptions) a somewhat mysterious paradigm of the *main root* (see e.g., [12], pp. 190, 224, 233 etc.) as a challenge for further investigations. Moreover, the power to compute efficiently roots of polynomials enables us to give global predictions at least in the so called *stable* model with quickness and precision. This was inconceivable some 60 years ago ([12], p. 237).

The method of *generating functions* – and working with their residuum poles – offers (in the context of this symposion for M.M. Richter) a bridge between *discrete* logical reasoning via linear recursions and an *approximate* estimation of their solutions via the poles of their generating functions. It should be mentioned, that there is an equivalent approach with matrices and their eigenvalues, which has been presented by P.H. Leslie (cf. [12], p. 227 ff.).

At the very beginning, I take from my shrine of mathematical saints the person between Al Khwarizmi and L. Euler, namely Leonardo of Pisa alias Fibonacci, the man, whose name is involved in more than a dozen different topics of modern theoretical informatics, even though he is not directly responsible for all this. We usually connect the association of "rabbits" with this name although we should better ascribe to him the glory of introducing the Indian-Arabic numbers into the western world – the basis of all serious calculation techniques in science. But our model starts indeed with Fibonacci's original contribution to rabbit-demography in his Liber Abbaci from 1202. We present a facsimile[1] of these very famous two pages, which is taken from a nice book of H. Lüneburg [6]. This author, who is a collegue of our celebrated jubilee from Kaiserslautern, coquets somewhat in not (or only scarcely) paying attention to the Fibonacci sequence F_n of rabbits at time n:

$$1, 2, 3, 5, 8, 13, 21, 34, 55, 89, 144, 233, 377$$

Does the abundance of many other important achievements of the Liber Abaci really make the rabbit business to a mere anecdote? It might enjoy the reader to read the latin text (with some help in punctuation):

Quot paria conicolorum in uno anno ex uno pario germinentur

Quidam posuit unum par cuniculorum in quodam loco, qui erat undique pariete
circumdatus, ut sciret quot ex eo paria germinarentur in uno anno: cum natura eorum
sit per singulum mensem aliud par germinare; et in s(ecundo) mense ab eorum natiui-
tate germinant. quia suprascriptum par in primo mense germinat, duplicabit
ipsum, erunt paria duo in uno mense. ex quibus unum s(cilicet) primum in secundo mense
(change of page)
geminat; et sic sunt in secundo mense paria 3, ex quibus in uno mense duo pregnantur;
et geminantur in tercio mense paria 2 coniculorum; et sic sunt paria 5 in ipso mense, ex quibus in ipso pregnantur paria 3; et sunt in quarto mense paria 8; ex quibus
paria 5 geminant alia paria 5. quibus additis cum parijs 8 faci-
unt paria 13 in quinto mense. ex quibus paria 5 quae geminata fuerunt in ipso
mense, n(on) c(on)cipiunt in ipso mense, s(ed) alia 8 paria p(re)gnantur; et sic s(un)t in sexto mense
paria 21. cum quibus additis parijs 12, quae geminantur in septimo, erunt in ipso
paria 34 cum quibus additis parijs 21, quae geminantur in octauo mense,
erunt in ipso paria 55; cum quibus additis parijs 34, quae geminantur in no-
no mense, erunt in ipso paria 89; cum quibus additis rursum parijs 55,
quae geminantur in decimo, erunt in ipso paria 144; cum quibus additis rursum
parijs 89, quae geminantur in undecimo mense, erunt in ipso paria 233.

[1] Biblioteca Nazionale Centrale Firenze, Codice Magliabechiano m.s. Conv. Soppr. C.I.2616, ff.: 123v, 124r. Reproduced by concession of the Ministero per i Beni e le Attività Culturali della Repubblica Italiana. No reproduction in any form without authorization by the Biblioteca Nazionale Centrale Firenze.

The inclusion of these facsimiles would not have been possible without the kind support of Heinz Lüneburg who provided the material.

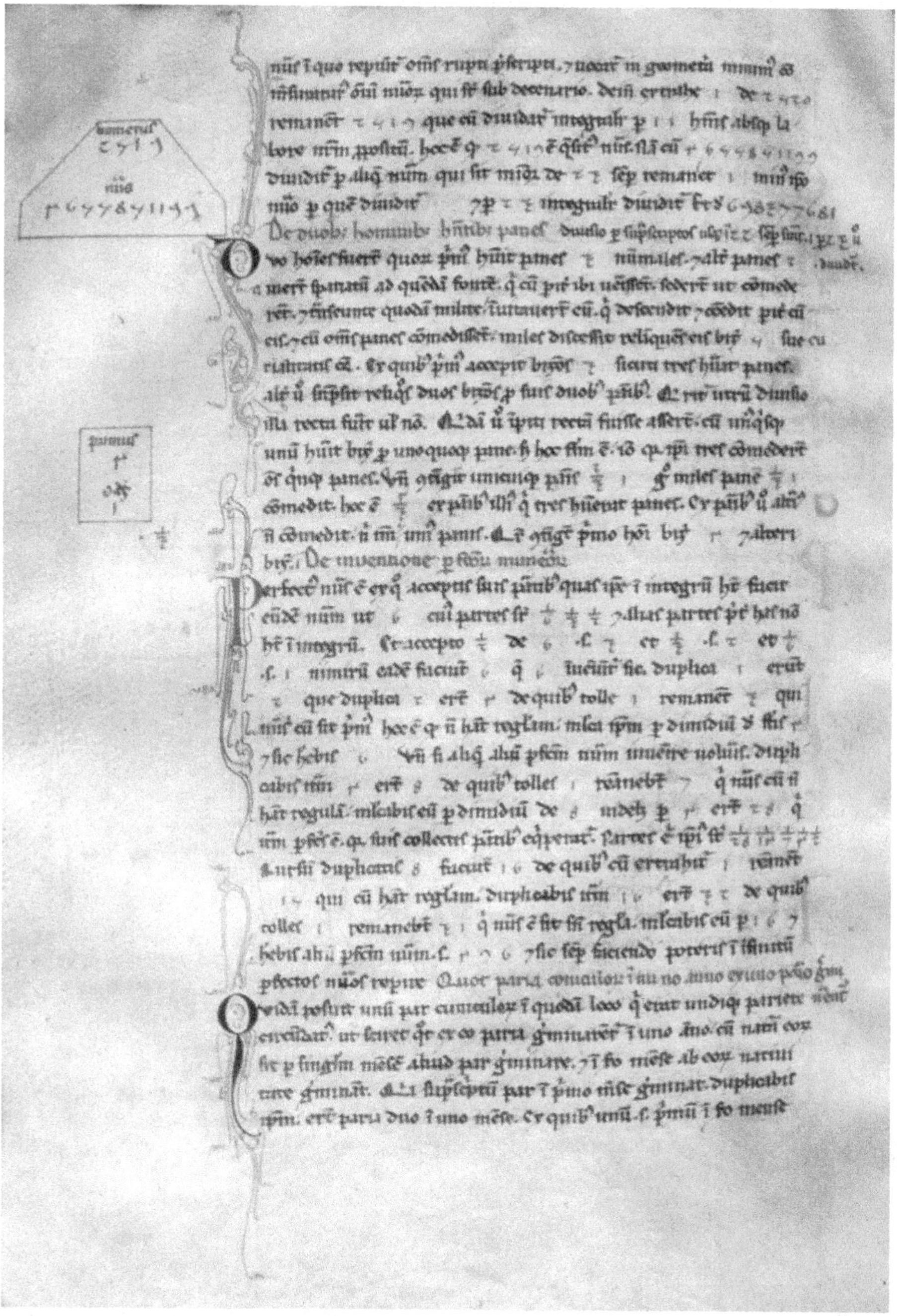

cum quibus et(iam) additis parijs 144, quae geminantur in ultima mense, erunt
paria 377. et tot paria peperit s(uprascrip)tum par in prefato loco in capite unius
anni. Potes enim videre in hac margine, qualiter hoc operati fuimus, s(cilicet) q(uod) iunximus
primum numerum cum secundo, uidelicet 1 cum 2; et secundum cum tercio; et tercium cum quarto / et quar-
tum cum quinto; et sic deinceps, donec iunximus decimum cum undecimo, uidelicet
144 cum 233. et h(ab)uimus s(uprascrip)torum cuniculorum summam, uidelicet 377.
et sic posses facere per ordinem de infinitis numeris mensibus

Note, that at the right margin of the facsimile the numbers are given again in a vertical table. It is curious, that the well renowned Fibonacci Association would not accept this text as a paper for its "Fibonacci Quarterly", since the Liber Abaci does not meet the Association's regulation to *begin* the sequence with

$$F_0 = 0, F_1 = 1, F_2 = 1, F_3 = 2, F_4 = 3, F_5 = 5, \ldots$$

We furthermore call attention to the fact, that Fibonacci's sequence has a strong connection to the golden section with the mysterious number $\Phi = 1.6180339\ldots = \frac{1}{2}(\sqrt{5}+1)$, a classical measure of architecture and art already in antiquity. It is furthermore amusing that Fibonacci's sequence seems to be a guideline in guessing future exchange rates by superstitious stock brokers (see, e.g., Frankfurter Allgemeine FAZ, Oct. 8, 2003).

The increase of rabbit population is in fact some crude form of demography: The *fertility* of the rabbits *never ends*, and moreover they are *immortal.* Therefore we extend this model and assume, that *fertility* starts at the age of f, that fertility *ends* at $g > f$ and that the individuum *dies* at age $h > g$. These assumptions seem still very simple, since according to this model each woman bears one daughter per time unit, since fertility ends simultaneously and since all women die at the same age. But under this "ideal" model the main root paradigm mentioned above will be preserved with some surprising facets and conjectures.

Finally we will take a look at a generalized application, which takes into account the average fertility of women at a certain age and which uses the mortality tables.

2 Generalized Fibonacci Demography

The connection between Φ and the Fibonacci sequence with its recursion $F_n = F_{n-1} + F_{n-2}$ is usually attributed to *Binet* (1843), but in fact *Euler* has given this formula already in 1765:

$$F_n = \frac{1}{\sqrt{5}}(\Phi^n - (1-\Phi)^n)$$

This result might be better understood by noting, that Φ and $(1-\Phi) = -0.6180339$ are the roots of the characteristic polynomial $1 + z - z^2$, which will be discussed later. The n-th powers of these roots – multiplied by appropriate cofactors (here $\frac{1}{\sqrt{5}}$ and $-\frac{1}{\sqrt{5}}$) – sum up to the numbers F_n. And the main root paradigm is indicated by the well known observation, that only $\frac{1}{\sqrt{5}}\Phi^n$ is crucial for the calculation of F_n, while the term $-\frac{1}{\sqrt{5}}(1-\Phi)^n$ serves only as a correction term with modulus always less than $\frac{1}{2}$. Therefore F_n is the next integer to $\frac{1}{\sqrt{5}}\Phi^n$. For $n = 10$ we have, e.g., according to this formula

$$F_{10} = 55.0036361\ldots - 0.0036361\ldots = 55.$$

Obviously, if fertility starts at time f, the recursion for F_n changes to $F_n = F_{n-1} + F_{n-f}$. But what happens, if the individuals *die* at the age $h \geq g$? We shall deal with this case in a moment.

Surprisingly the recursion $F_n = F_{n-1} + F_{n-f} - F_{n-h}$ does *not* take care of this situation (as some authors have erroneously asserted), since then individuals would die several times. Instead, this recursion describes the situation, that the individuals become *non-fertile* at this time: Of course, the fact, that an individual of age g is non-fertile can be interpreted as giving at each time birth to a new object, which dies immediately. And what about mortality? Let $F_n^{(f,g)}$ be the total number of immortal individuals with childhood of length f and infertility after g units. If they die at time $h \geq g$, then the total number of living individuals at time n is obviously $T_n^{(f,g,h)} = F_n^{(f,g)} - F_{n-h}^{(f,g)}$. It is therefore sufficient to analyze $F_n^{(f,g)}$ in order to get full information, and the mathematical significance of h drops out in a rather trivial manner.

3 The Generating Functions

It can be shown with standard arguments, that the generating function (GF) $F^{(f,g)}(z) = \sum_n F_n^{(f,g)} z^n$ is given by $\frac{z}{1-z-z^f+z^g}$.

The classical case is $f = 2$ and $g = \infty$ (i.e. z^g disappears).

Proof sketch: The sequence $F_n^{(f,g)}$ starts at $F_1^{(f,g)} = 1$ and then goes back to the earlier values according to the recursion $F_n^{(f,g)} = F_{n-1}^{(f,g)} + F_{n-f}^{(f,g)} - F_{n-g}^{(f,g)}$.

Thus in the case $f = 4, g = 10$, e.g., we get the sequence

$$1, 1, 1, 1, 2, 3, 4, 5, 7, 10, 13, 17, 23, \ldots$$

Under this scenario

$$(1 - z - z^4 + z^{10})(z + z^2 + z^3 + z^4 + 2z^5 + 3z^6 + 4z^7 + 5z^8 + 7z^9 \\ +10z^{10} + 13z^{11} + 17z^{12} + 23z^{13} + \ldots) = z$$

and in the same way also in general the GF is $\frac{z}{1-z-z^f+z^g}$. One gets the nominator of the GF for a linear recursion with constant coefficients by multiplying the beginning of $F_n^{(f,g)}$ – which is given by the initial values – with the polynomial $g^{(f,g)}(z)$ which is assigned to the recursion. The denominator polynomial

$g^{(f,g)}(z)$ is called the *auxiliary* polynomial of the recursion – it is the *reflexion* of the better known *characteristic* polynomial $g_{(f,g)}(z)$, i.e. it is obtained from $g_{(f,g)}(z)$ by substituting z by $\frac{1}{z}$ and then multiplying with z^g (cf.e.g. [9], p. 84). Obviously the roots of both types of polynomials are inverse to each other. Unfortunately auxiliary and characteristic polynomials are often confused.

Now *Binet's formula* can be *generalized* for sequences F_n , which are defined by arbitrary linear recursions with constant coefficients. The corresponding *GF*'s are exactly the rational functions $F(z) = \frac{f(z)}{g(z)}$. If all the poles of $F(z)$ are *simple*, an explicit formula for F_n is well known: $F_n = \sum_i c_i \frac{1}{a_i^n}$, where the a_i are the roots of $g(z)$.

The ***cofactor*** c_i can be determined from the partial fraction (Laurent contribution) $T_i = \frac{b_i}{z-a_i}$ by putting $c_i = -\frac{b_i}{a_i}$ such that $\frac{c_i}{1-\frac{z}{a_i}}$ is the i-th contribution to the Binet formula.

If a_i is a *k-fold* root ($k > 1$), then from the corresponding principal part in the Laurent expansion of $F(z)$ at a_i the cofactors appear in the form

$$T_i = \frac{c_{i_1}}{1-\frac{z}{a_i}} + \frac{c_{i_2}}{(1-\frac{z}{a_i})^2} + \ldots + \frac{c_{i_k}}{(1-\frac{z}{a_i})^k},$$

and using the well known identity

$$\frac{1}{(1-\frac{z}{a})^j} = \sum_n \binom{n+j-1}{j-1} \left(\frac{z}{a}\right)^n$$

one then gets corresponding contributions to the Binet formula in the general case. There is a strict analogy to corresponding facts with linear differential equations.

For investigations with generating functions one needs comfortable formulas for the Laurent coefficients. In order to avoid calculations via systems of linear equations one can use for *simple* roots the well known formula $b_i = \frac{f(a_i)}{g'(a_i)}$. Explicit formulas in the case of k-fold roots are not so well known, but can be given in a rather simple way with the help of Bell polynomials [8].

From Binet's general formula it is clear, that the most important contribution to $F_n^{(f,g)}$ from roots of the auxiliary polynomial will have modulus < 1.

4 Perpetual Fertility and the Roots

We first investigate the case $g = \infty$, i.e. neither mortality nor non-fertility at all. The *classical* case $f = 2$ has been analyzed throughout.

The *general* polynomials $g^{(f,\infty)}(z) = 1 - z - z^f$ have been investigated by K. Dilcher [3] in their reflected form $z^f - z^{f-1} - 1$. Dilcher's results say roughly (in our terminology), that there is a distinguished single real root $a_1^{(f,\infty)}$ in $0 < a_1^{(f,\infty)} < 1$, which is responsible for the main term in the Binet formula for $F_n^{(f,\infty)}$. Only one more real root $a_2^{(f,\infty)}$ (which is < -1) may exist: This happens if f is even. Roughly one third of the (conjugate) complex roots of $g^{(f,\infty)}$ have modulus less than 1, and all roots – written in the polar form $z = re^{i\Theta}$ — are somewhat uniformly distributed with respect to their argument Θ in an annular region around the unit circle. Furthermore r is growing monotonously with Θ.

We add here only some few comments:

Since the polynomial $g^{(f,\infty)}$ is very similar to a cyclotomic polynomial, it is not too surprising, that a similar behavior of the roots appears. Only the occurrence of the main root is an additional feature.

Furthermore: Since for zeros of $1 - z - z^f$ both real and imaginary parts vanish, from

$$r^f \cos f\Theta = 1 - r\cos\Theta \text{ and } r^f \sin f\Theta = r\sin\Theta$$

we get by squaring and adding both equations

$$1 - 2r\cos\Theta + r^2 - r^{2f} = 0$$

as the equation of a curve in the polar plane, on which the zeros are situated: We can isolate $\cos\Theta$ and find, that for all r the relation between $\cos\Theta$ and r is one-one and continuous. For $r = 1$ we get $\cos\Theta = \frac{1}{2}$, i.e. $\Theta = \pm\frac{1}{3}\pi$. Therefore – independent from f — exactly one third of the full circle area around $z = 0$ with the positive real axis as symmetry contains zeros which have modulus ≤ 1.Thus there will be a (small) fraction of those roots, which contribute an eventually (for $n \to \infty$) non-vanishing term to the Binet-formula.

We mention, that for similar problems in the Fibonacci-context $f = 2$ roots of the denominator polynomials have been investigated by several people (e.g. [4], 270/71).

In all our considerations the polynomials have the absolute term ± 1. It is to be expected from Vieta's root theorem, that the existence of a distinguished main root a_1 with $|a_1| < 1$ must be compensated by the fact, that *other* roots with modulus < 1 are somehow in a minority against the roots with modulus > 1.

Obviously the main root decides the asymptotic order of magnitude for the population numbers, since all other roots share only an eventually vanishing *relative* part of these quantities. But in order to show, that the main term is for large n *absolutely* precise (i.e. that the exact values are got by simply rounding the main term, because the error has modulus less than $\frac{1}{2}$), one would have to show, that *all* other roots have modulus > 1. This is according to Dilcher in general *not* the case for the numbers $F_n^{(f,\infty)}$. Thus we can only save a weak version of the main root paradigm into the case of arbitrary f.

5 Loss of Fertility at Finite Age

We shall investigate now, what happens, if non-fertility occurs at the finite time $g > f$. Here we have to investigate the roots of the auxiliary polynomial $g^{(f,g)}(z) = 1 - z - z^f + z^g$. We start with the classical case $f = 2$.

For fixed n and $g \geq n$ we have the Fibonacci numbers $F_n^{(2,g)} = F_n$, since loss of fertility doesn't have influence in this initial case.

The polynomial $g^{(2,g)}(z) = 1 - z - z^2 + z^g$ has a simple real root $z_1 = 1$ and in addition, if m is odd and $g > 3$, another one at $z_g = -1$. More important, there is for $g > 3$ a real root z_2 in $\Phi = \frac{1}{2}(\sqrt{5} - 1) = 0.61803\ldots < z_2 < 1$, which we call the *main* root and which tends to Φ if $g \to \infty$.

Theorem 1. *All non-real roots of* $g^{(2,g)}(z) = 1 - z - z^2 + z^g$ *(occur in conjugate complex pairs and) have modulus* > 1.

Proof: For a zero $z = re^{i\Theta}$ of $1 - z - z^2 + z^g$ both real and imaginary part vanish. We use the conjugate complexity of roots and consider only the upper halfplane $0 \leq \Theta \leq \pi$. Then

$$r^g \cos g\Theta = r^2 \cos 2\Theta + r\cos\Theta - 1; \; r^g \sin g\Theta = r^2 \sin 2\Theta + r\sin\Theta.$$

We abbreviate $C_k := \cos k\Theta$; $S_k := \sin k\Theta$. $C := C_1$; $S := S_1$. On squaring and adding both sides we get $r^{2g} = r^4 + r^2 + 1 + 2r^3(CC_2 + SS_2) - 2r^2C_2 - 2rC$.

Since $C_2 = C^2 - S^2 = 2C^2 - 1$ and $S_2 = 2SC$, we have

$$1 + 3r^2 + r^4 - r^{2g} + 2r(r^2 - 1)C - 4r^2C^2 = 0.$$

This relation represents a continuous curve in the polar (r, Θ)-plane with a simple connection pattern, since the equation is only quadratic in C. Only for $\Theta = 0$, i.e. $C = 1$, we have an isolated region with a root $(r, 0)$ where $0 < r < 1$, and r fulfils the equation $1 - 2r - r^2 + 2r^3 + r^4 - r^{2g} = 0$. On the other hand, putting $r = 1$, we get $C^2 = 1$ or $C = \pm 1$. Using continuity it is easy to see, that the zeros with Θ in $0 < \Theta < \pi$ have either *all* $r > 1$ or *all* have $r < 1$, and it is immediate, that the first alternative is correct. Therefore, the principal root z_2 yields the main contribution in the Binet formula for $F_n^{(f,g)}$, and – except bounded contributions which result from z_1 and z_g – the contributions of the complex roots tend to zero if $n \to \infty$.

In the general case $f \geq 2$ the situation is more interesting:

A region $\Re^{(f,g)}$ (root region), which contains (among other points) the zeros of $g^{(f,g)}(z)$ in the upper half plane, is now defined by

$$r^{2g} = r^{2f} + r^2 + 1 + 2r^{f+1}(CC_f + SS_f) - 2r^f C_f - 2rC$$

Putting $r = 1$ we get $CC_f + SS_f - C_f - C + 1 = 0$.

We require more trigonometry and use, that C_f can be written as polynomial in C using the Chebyshev polynomials T_n: We have $C_f = T_f(C)$ (cf. [1], 22.3.15 and p. 795). In addition it can be proved that $C_f = CC_{f+1} + SS_{f+1}$. Thus we have for the zeros with modulus $r = 1$

$$C_f - C_{f-1} + C - 1 = 0$$

Using these relations we see, that the region $\Re^{(f,g)}$ in the (r, Θ)-plane is such, that putting $r = 1$ we get f zeros for the resulting polynomial in C. We call these zeros *nodes* of the root region. In general nodes ar *not* zeros of $g^{(f,g)}(z)$, since the root regions contain many other points, which are not zeros of the polynomial. But nodes are relevant for judging, whether zeros of $g^{(f,g)}(z)$ change their property of having modulus larger or smaller than 1. Our result is now the

Theorem 2 (Node Theorem).
The node equation *is* $N_f(z) = T_f(z) - T_{f-1}(z) + z - 1 = 0$. $N_f(z)$ *is* independent from g.

Thus for fixed f and different g the root regions are winding in a certain way around the unit circle, but they all have the *same* nodes.

Knowing this and encouraged by corresponding properties of the Chebyshev polynomials one can determine the nodes of $N_f(z)$ exactly:

Theorem 3. *$N_f(z)$ has f simple real roots in $-1 \leq z \leq +1$, and these are situated at the angels $v_j = 2j\frac{180}{f}$ for $0 \leq j \leq \frac{f}{2}$ and $w_j = (1+2j)\frac{180}{f-1}$ for $0 \leq j < \frac{f}{2}$.*

The proof is by verification and uses the fundamental identity $C_f = T_f(C)$.

As an illustration, $N_{10}(z)$ has its nodes at the angles 0°, 36°, 72°, 108°, 144° and at 20°, 60°, 100°, 140°, 180°.

We call a Θ-region between two consecutive nodes ***relevant***, if the root region is in the *inner* part of the unit circle. The zeros z_i of $g^{(f,g)}(z)$, which lie on these parts, contribute essentially to the Binet formula for $F_n^{(f,g)}$, since $\left|\frac{1}{z_i}\right| > 1$.

In the example $f = 10$ the relevant regions belong to the Θ-intervals $I_1 = [20°; 36°]$, $I_2 = [60°; 72°]$, $I_3 = [100°; 108°]$ and $I_4 = [140°; 144°]$. The total length of these intervals in degree measure is $\lambda_{10} = 40°$. It is an easy exercise to prove, that for $f \to \infty$ λ_f tends to 45°. More detailed for f even $\lambda_f = 45\frac{f-2}{f-1}$ degrees, for f odd $\lambda_f = 45\frac{f-1}{f}$ degrees. The special case $f = 2$ yields, as we saw before, $\lambda_2 = 0$. Note, that the root region in the previous section, which yielded the node with $\Theta = \frac{1}{3}\pi$, i.e. 60°, is different from the root regions $\Re^{(f,g)}$ under consideration at present: The former root region was defined under the special assumption $g = \infty$. Of course, there are always many nontrivial possibilities to define root regions for a polynomial.

We conjecture, that in the limit $f \to \infty$ one quarter of the angle region produces relevant zeros of $g^{(f,g)}(z)$. All this is of course very plausible, if we remember the remark on Vieta's theorem. Numerical evidence supports another conjecture, that – similar to the case $g = \infty$ – the roots of $g^{(f,g)}(z)$ are somewhat uniformly distributed over the Θ-region. For $f = 10$ and $g = 100$, e.g., we have in the upper halfplane 4 roots in I_1, 3 roots in I_2, 2 roots in I_3 and one root in I_4, while the remaining 39 nontrivial complex roots are not relevant.

There is one disappointing feature in the general case, which seems to destroy the nice dream of a clean main root paradigm: A detailed inspection shows, that the root regions are *no* longer *connected* curves: Certainly, $\Re^{(f,g)}$ *contains* a continuous curve, which is winding around the nodes. But there may exist separate branches of $\Re^{(f,g)}$ – *bubbles*, which may also contain roots of $g^{(f,g)}$.

These bubbles are often tiny and cannot be found by usual computer algebra programs, since a command to plot the root region of a relation, which is defined by a high-degree polynomials, is normally not available, or precision is too low, respectively. The background of this observation is, that the node equation is of high degree in C. In the example $f = 10$ and $g = 100$ there is – besides the main root $r = 0.835079$ – in the upper halfplane one exceptional zero with $r = 0.931488$ and $\Theta = 30.692°$ in the relevant region I_1, which is not on the main branch of $\Re^{(f,g)}$. All other zeros have modulus > 0.9852. This unexpected feature throws light on another fact: The distinguished real main root seems no longer look that amazing: The main root is simply also situated on a bubble of $\Re^{(f,g)}$.

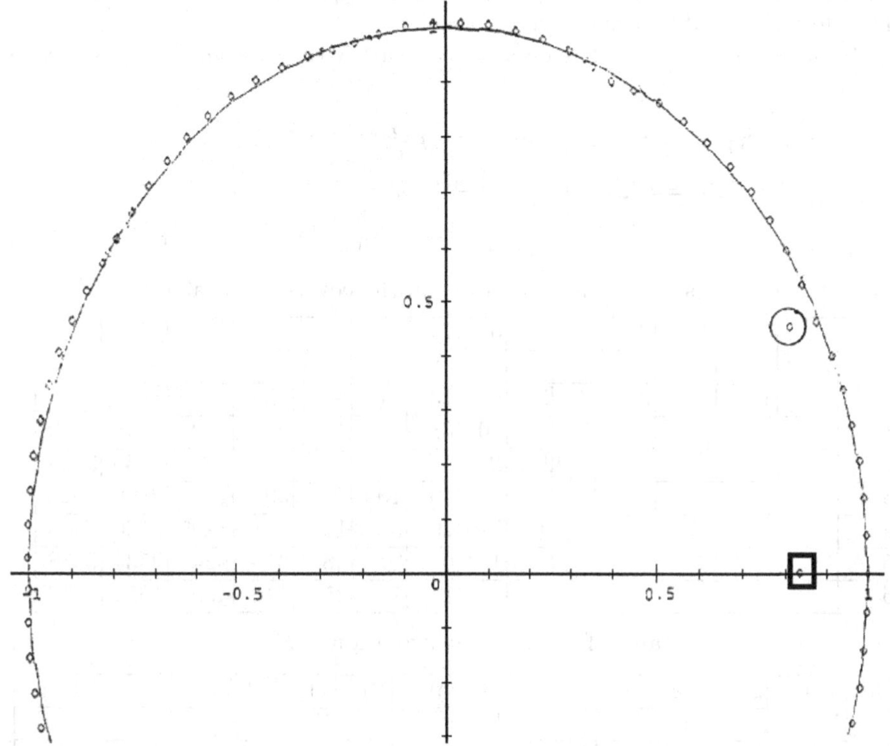

Fig. 1. Roots of $1 - z - z^{\cdot\cdot} + z^{\cdot\cdot\cdot}$ in the upper halfplane. The two exceptional roots are emphasized

6 The Convolution Method

In contrast to the Binet formula there is a totally different way to calculate the quantities $F_n^{(f,g)}$. Such a new method seems desirable, since the main root paradigm becomes suspect for $g \neq \infty$. Therefore we discuss a second way to estimate the growth of $F_n^{(f,g)}$ asymptotically. We start with the values $F_n^{(f,\infty)}$

from section 4 and consider them now as well known. These numbers work as first approximations for the numbers $F_n^{(f,g)}$.

Proposition 1. *The difference* $F_n^{(f,\infty)} - F_n^{(f,g)}$ *for* $n > g$ *can be calculated by using the entries from the* convolution *tables induced by the numbers* $F_n^{(f,\infty)}$.

As a preparation, we define ${}^{(i)}F_{(f,\infty)}(z) = \left(\frac{z}{1-z-z^f}\right)^{i+1}$ as the i-th convolution of $F_{(f,\infty)}(z) = \frac{z}{1-z-z^f}$. For $i = 0$ we have the original sequence $F_n^{(f,g)}$. The convolution tables can be constructed easily: Starting with the trivial first line, which has formally $i = -1$, the next lines can be produced from left to right adding an entry to the left of the actual place with the upper left number according to an obvious "template" (for the simplest case $i = 1$ cf. e.g. [10]). Therefore these tables are known in principle.

In order to illustrate what happens we start with two somewhat mysterious examples:

$$F_{20}^{(2,8)}(= 5479) = 6765 - 1308 + 22 = F_{20}^{(2,\infty)} - {}^{(1)}F_{13}^{(2,\infty)} + {}^{(2)}F_{6}^{(2,\infty)}$$

$$\begin{aligned} F_{20}^{(3,6)}(= 223) &= 872 - 816 + 171 - 4 + 0 = \\ &= F_{20}^{(3,\infty)} - {}^{(1)}F_{15}^{(3,\infty)} + {}^{(2)}F_{10}^{(3,\infty)} - {}^{(3)}F_{5}^{(3,\infty)} + {}^{(4)}F_{0}^{(3,\infty)} \end{aligned}$$

The r.h.s.-numbers are significant entries in the convolution tables:

i\n	0	1	2	3	4	5	6	7	8	9	10	11	12	13
	1													
0		1	1	2	3	5	8	13	21	34	55	89	144	233
1			1	2	5	10	20	38	71	130	235	420	744	**1308**
2				1	3	9	**22**	51	111	233	474	942	1836	3522
3					1	4	14	40	105	256	594	1324	2860	6020
4						1	5	20	65	190	511	1295	3130	7285
5							1	6	27	98	315	924	2534	6588
6								1	7	35	140	490	1554	4578

Table of Fibonacci convolution ${}^{(i)}F_n^{(2,\infty)}$

i\n	0	1	2	3	4	5	6	7	8	9	10	11	12	13	14	15
	1															
0		1	1	1	2	3	4	6	9	13	19	28	41	60	88	129
1			1	2	3	6	11	18	30	50	81	130	208	330	520	**816**
2				1	3	6	13	27	51	94	**171**	303	527	906	1539	2586
3					1	**4**	10	24	55	116	234	460	879	1640	3006	5424
4						1	5	15	40	100	231	505	1065	2175	4320	8391
5							1	6	21	62	168	420	987	2220	4815	10122
6								1	7	28	91	266	714	1792	4278	9807
7									1	8	36	128	402	1152	3072	7752
8										1	9	45	174	585	1782	5028

Table of convolution with $f = 3$

Theorem 4 (Shift Theorem).

$$F_n^{(f,g)} = \sum_i (-1)^i \, {}^{(i)}F_{n-i(g-1)}^{(f,\infty)}.$$

$F_n^{(f,g)}$ can be represented by a quickly descending sum of convolution numbers with alternating signs.

The proof uses generating functions:

$$\begin{aligned} F^{(f,g)}(z) &= \frac{z}{1-z-z^f+z^g} = \frac{z}{1-z-z^f}\frac{1}{1+\frac{z^g}{1-z-z^f}} \\ &= \frac{z}{1-z-z^f}\sum_i\left(-\frac{z^g}{1-z-z^f}\right)^i \\ &= \sum_i (-1)^i (z^{g-1})^i \, {}^{(i)}F^{(f,\infty)}(z), \end{aligned}$$

and this was the claim.

An asymptotic estimate of the $(i-1)$-fold convolution numbers follows from the partial fractions for the generating functions ${}^{(i)}F^{(f,\infty)}(z) = (\frac{z}{1-z-z^f})^i$, which can be calculated using the techniques for i-fold roots (mentioned in section 3).

The values for the convolution numbers in the Fibonacci case $f = 2$ are included in the sequences M1377, M2789 and M3476 of [11].

7 Application

In order to extend the model to realistic conditions one has to take into account the different ages, at which women bear girls. It is a straightforward task to integrate those detailed statistical data into our previous considerations, which used *uniform* data for fertility and its loss. Diversified mortality statistics for women can also be integrated into this framework. Under these assumptions we report on an experiment, for which we assumed, that (in Germany) women produce during their lifetime 0.7 girls on the average. We paid attention to the different ages, when mothers bear their daughters and to the (slightly simplified) mortality statistics for women. We have compressed data to five-year-periods. Starting with an initial cohort of 100000 women we found the following table:

years	#women	years	#women	years	#women
2	100000	87	174045	172	51178
7	100000	92	158740	177	47685
12	100000	97	146791	182	44416
17	105000	102	141208	187	41328
22	126000	107	132176	192	38442
27	151000	112	121631	197	35775
32	164250	117	111874	202	33310
37	171350	122	104221	207	31013
42	179260	127	97901	212	28864
47	191073	132	91464	217	26859
52	201440	137	84824	222	24998
57	208989	142	78583	227	23270
62	213198	147	73135	232	21661
67	**214010**	152	68283	237	20161
72	213034	157	63665	242	18764
77	206424	162	59193	247	17464
82	192253	167	54997	252	16256

The auxiliary polynomial has in this situation a clearly separated single real main root $a_1 = 1.074382163$, which induces an ultimate shrinkage by 6.9 percent within five years. This means an annual decrease of about 1.42 percent. It can be argued, that the ultimate rate of decrease does not depend on mortality statistics – which we assumed to be fixed. Only the cofactor to $\frac{1}{z-a_1}$ may change with an increase or decrease of life expectation.

The table confirms the heuristic expectation, that the initial cohort together with its female descendants attains its maximum after 70 years and goes down to one third after two centuries.

8 Conclusion

We have demonstrated, that a discrete population model can be a serious instrument of demography. The investigation of the denominator-roots of the generating functions gives direct hints for the future population. It may be permitted to dream of a computing technology, which admits to visualize the future development instantly, when the values of the demographic parameters are dynamically and quasi-continuously changed. If one would be able to calculate the roots of a high-degree polynomial in real time, the present paper could be a help to realize this utopia.

References

[1] Abramowitz, M., Stegun, I.A.: Handbook of Mathematical functions, Dover Publ. Co. New York 1965
[2] Comtet, L.: Advanced Combinatorics, D.Reidel 1974
[3] Dilcher, K.: On a class of iterative recurrence relations. In "Applications of Fibonacci Numbers", C.E. Bergum et al. (eds.) 5 (1993), 143–158, Kluwer Acad. Publ.
[4] Dubeau, F.: The rabbit problem revisited, Fibonacci Quart. 31 (1993), 268–274
[5] Graham, R.L., Knuth, D.E., Patashnik, O.: Concrete Mathematics, Addison-Wesley 1989 (2nd ed.)
[6] Lüneburg, H.: Leonardi Pisani Liber Abaci oder Lesevergnügen eines Mathematikers, BI Wissenschaftsverlag (2nd ed.), Mannheim etc, 1993
[7] Mueller, U.: Bevölkerungsstatistik und Bevölkerungsdynamik, de Gruyter, Berlin and New York 1993
[8] Oberschelp, W.: Unpublished note 2001. A copy can be supplied on demand
[9] Ostrowski, A.M.: Solutions of Equations in Euclidean and Banach Spaces, Academic Press , New York 1973
[10] Riordan, J.: Combinatorial Identities, Wiley 1968
[11] Sloane, N.J.A., Plouffe, S.: The Encyclopedia of Integer Sequences, Academic Press 1995
[12] Smith, D., Keyfitz, N. (eds.): Mathematical Demography, Springer, Berlin etc. 1977
[13] Wilf, H.: Generating functionology, Academic Press 1990 (2nd ed.)

Computer Science between Symbolic Representation and Open Construction

Britta Schinzel

Institute for Computer Science and Society of the University of Freiburg
schinzel@modell.iig.uni-freiburg.de

Abstract. Computer science (and AI along with it) has fundamentally different operational possibilities. Firstly, in that humans represent a problem area explicitly symbolically and put the solution of the problem into algorithm, in order to ensure a complete problem solution. Secondly, in that they – using the preconditioned computer less as a controlled transformation medium than as a to a certain extent unknown physical system – initialise a certain approach and observe the calculation process, and thirdly as a medium for the representation of pictures, dynamics, etc.. My paper focuses on basic questions of representation by means of computers, which are directed in particular towards the character of the symbolic and the pictorial, the discrete and the continuous, and thinking in symbols and in analogue structures respectively.

The opposition of logic and approximation is connected with the oppositions of discrete and continuous, that is finite and infinite, digital and analogue, number/script and picture and also calculation and simulation, closed and open solutions, structuralism and constructivism. Signs, symbols, numbers, letters, letter scripts (such as the Korean or Roman alphabets), algorithms, logic, complete models are therefore opposed to the continuous, the analogue, pictures, pictorial script (Chinese, some Japanese scripts), simulation, statistics and probability, and evolutionary models.

1 Characteristics of Computer Science as a Science

A science is generally characterised by its subject, theory/ies and method/s. In addition however, it is also determined by its epistemological aim/s and purpose/s, those of its findings and the nature of the research processes themselves, such as laboratory or field studies, the intellectual work of mathematicians or the computer work, not only of computer scientists, that is by the historically developing processes which guide research. In the case of computer science, it is striking that it is difficult to determine the subject and epistemological aims and purposes. That distinguishes it from the traditional technical sciences, which have a well-defined subject area to which basic knowledge can be applied, where material and task determine the method, and whose subjects are conversely limited by their methodical accessibility (Pflüger 1994). Although the computer is a subject of computer science research and is the subject and the material (in the engineering sense), with which computer science

W. Lenski (Ed.): Logic versus Approximation, LNCS 3075, pp. 59–76, 2004.

operates[1], it does not take on the role that for example the object "living material" plays in biology. To a certain extent, the applications of computer science play this role as starting material for focussing on computer-technical realisation and problem-solving. However these application areas are almost arbitrary in computer science, and they are primarily the subjects of other sciences, such as biology for example. Pflüger therefore refers to a missing subject reference, which he claims makes computer science boundless and unfounded. "The method is not mediated by the subject, is no longer the way which the thing itself goes, but obeys a scheme of the machine, which is pulled on to cover reality. Computer science operates in very different subject areas; social, scientific and technical, with one homologue approach, that of computer programming."

Computer scientific problem-solving, for biology for example, has an effect on both biology and computer science. According to Pflüger, computer science within a particular context leads to "unfoundedness" towards the context, to the phenomenon that applications of computer science are able to fundamentally influence and change their methods and even their medium, the computer. To an increasing extent, the contexts determine findings interests and methods of computer science, and even change the medium of the computer itself, whose architecture ranges from the von Neumann computer to parallel, distributed or connectionistic systems, right up to the material, if we consider alternative materials such as biological or atomic chips instead of silicon-dioxide chips in genetic and DNA computing or in quantum computers. On the level of representations, we can observe a movement from language to pictures, from the transformation of symbolic representations to that of signals, impulses or parameterised data without direct meaning for humans, from closed to open systems.

At this point the question of the constants of computer science also arises. To what extent formalisation and symbolic representation[2] can still be used to describe computing methods, must be discussed in connection with computing representations. We must distinguish between formalisation and symbol processing on one side and signal and code processing on the other side, in processing data, numbers, signs, vectors or general complex data. One means of differentiating can be, for example, the place where meaning appears for humans, on the level of atomistic signs or on the holistic level. The replacement of the von Neumann computer with parallel and evolutionary hardware systems further results in shifts from linearity to multi-dimensional, parallel or holistic processing of signals and data. The importance of algorithms is thereby reduced in favour of simulating and evolutive processing methods, for which empirical observation is gaining importance as opposed to verification, and thereby the open construction with regard to results and meaning. A further important element[3] of epistemic changes, along with the visualisation of dynamics, is the increase in pictorial as opposed to textual representation made possible by increasing memory space.

---

1 Software is rather still an object of computer science, but as it only exists in connection with hardware and cannot be principally separated from hardware – every software can also be made as hardware – we may simply refer to computers.

2 Semiotics is often used in an attempt to find a methodical foundation for computer science, cf. e.g. Andersen, 1990; Nake, 1993, pp. 165 – 191.

3 A second, connected element is the increase of simulations, from open as opposed to closed provable solutions, whereby computer science also comes closer to sciences.

2 The Opposition of Discrete and Analogue

Discreteness and continuum are models of reality which describe reality differently and can be applied differently. In order to make use of them in the context of computer science, the opposition of logic and approximation must first be referred back to the differentiation between representation and the matter represented, between model and reality. Then discrete or analogue representations and/or their approximations and their transformations can be chosen accordingly. Finally, abstract symbolic representations, and also continuous representations, can be distributed in analogue form for adaptation to the extensively analogue human cognitive orientations. In differentiating between reality and physical measurement of this reality, that which is measured, which can usually be analogue (acoustic signals), but also discrete (measured data in scintillation counters, bubble chamber or in an MRI cylinder) we can already see that the difference between analogue and discrete is not an absolute difference. This is visible for example in photography, where the dualism of wave and corpuscle, or in rather more macroscopic terms, picture on photo paper and grain of silver bromide, makes the difference dependent on the "zoom", on the sharpness of the observation. After analogue measurement, conversions of the measured item (e.g. analogue-to-digital converter) can support further manipulation and processing into diverse forms of representation (e.g. MRI measurement values are transformed back into Maxwell equations to describe areas and these are further processed for digital image production).

Relationship between Model and Reality in Computer Science
Explicitly representative computer modelling usually follows the *copy perspective*, not the *construction perspective*, i.e. the cognitive process, which is intended to produce a relationship between a formal model and the assumed reality and as a result then sets off the software production and construction processes, observes the *position of the Copy or Correspondence Theory* (Tarski 1983). For the epistemological relation between model and original, it is assumed that the reality in which we live exists as objectively given, independent of cognitive models. Should it be successfully reproduced in a model, this model is a reproduction of an original. In contrast, objections against this can be adopted in variously extreme relativisations of the constitution of reality and its epistemological possibilities, e.g. with various radical *constructivist positions*: even assuming there is a reality independent of our perception, we interpret what we perceive hermeneutically according to our experiences and prejudices, wishes and objectives, in a historically and culturally contingent manner. Computer scientists also often take a *pragmatic position*. The question of whether our models reproduce an objective reality or whether they construct this in the first place is the wrong question: a model is not good because it is a true copy of reality, but because we can successfully operate with it in reality. The ontological question is therefore replaced by the relation of the means to the purpose. However such a model marks that which is perceived or specified with a special order, it can omit areas of reality which appear unimportant, including those that prove to be essential in hindsight through their functional omission (incomplete specification). This necessarily represents the specific and restricted view of the section of reality of those involved in the modelling (determination of requirements) with regard to the relationising and validation. To what extent this causes damage in the context in which the prob-

lem-solving is embedded depends on the adequacy of this view for the processes in the individual context.

Other models not based on explicit symbolic reconstruction can simulate input-analogue physical processes in analogue or hybrid computers for example, they can further process discrete input data such as image data like in a Boltzmann machine, or they can leave the problem of fitting together the suitable DNA strands to the DNA soup as in DNA computing.

Discrete / Analogue

A symbolic representation is always discrete, even if it should stand for something continuous. The approximation also has a discrete character, i.e. it remains in discrete representation, unlike interpolation, but like the latter it connects discrete and continuous areas, in this case in that - with the human perception of reality – that which is represented, the continuous, the infinite is approximated by something discrete, in that case the representation itself stands for a mathematical constant. Logic like the typography comes from symbolism as well.

Analogue representations are (separately) continuous descriptions, represented as curves or as mathematical formulae or as reproducible physical recordings, which vary proportionally to reality. An acoustic signal for example can be recorded by physical measuring equipment in a diagram in wave form on various media (tape, record, even an oscillator), varying in correspondence with the real sound, capturing this occurrence or only varying temporarily with it. A photograph varies shades of grey with that which is photographed. Such an analogue recording possibly contains more detailed information than a digital recording, however it is far less precisely storable and transferable than the digital recording. Analogue recordings cannot be copied or manipulated arbitrarily without changing the sound, increasing hissing, reducing precision. Analogue storage media therefore do not store precise measurements but intervals, between which the measured signal can no longer be exactly determined.

In computers, analogue and discrete elements are connected on various levels. On the lowest material level everything is analogue, the various hardware states are analogue but can also be immediately reinterpreted discretely, the state transformations as electron conversions have an analogue character. Discrete elements can be seen in the symbolic code, in assembler or programming languages for producing these state transformations, and in the symbolic representations, models and specifications, both as a means of interpersonal understanding and as a means of preparation for the programming, and as a language in themselves.

Analogue-to-Digital Converters [4]

The technology of A/D-conversion from analogue to digital takes measurements made by technical sensors in the environment as its starting point: changes in light, gravity, temperature, moisture, noise or pressure are transformed into voltage when the change in resistance at the measuring device creates a change in electrical voltage. This appears as a mathematical equation or in an oscillograph in the form of a con-

---

[4] http://wwwex.physik.uni-ulm.de/lehre/PhysikalischeElektronik/Phys_Elektr/node129.html, http://home.t-online.de/home/u.haehnle/technik/ad-wandlung.htm

tinuous curve, for single occurrences in impulse form for example, for complex occurrences as a frequency image. The converter converts single impulses into rectangular impulses, complex frequency images are dismantled into components (e.g. in sinus form) and represented after analysis through characterising staircase functions. Various processes, such as the tracking process, weighing process, sigma-delta process and one/two ramps process build up a reference voltage over equidistant voltage values and reduce this again in tacted impulses, thus creating the aforementioned staircase functions, from which number values can be read. Practical and amplification problems can also occur with these counting techniques, even if precision can be improved by investing a greater amount of time.

Digital technology converts analogue signals into numbers. The conversion process itself, however, harbours similar precision losses to the recording of analogue signals. Once the signals have been digitalised the inexactitude is reduced (without disappearing altogether), as the copying and transferral process is essentially made up of the repetition of these numbers. Were it possible to store numbers as such directly, the ideal situation of potential flawlessness would be achieved, as numbers are symbols, non-physical entities. Unfortunately however, numbers cannot be stored, but only their physical forms of appearance, that is signals for digital entities. Numbers have a material form only as signals, and all operations of storing, transferring and copying are only as precise as the signal technology that lies beneath them – although whole numbers can be restored within the precision boundaries of correcting codes and returned to their original value. In as far as these corrections function flawlessly, digital signals can be stored, transferred and copied without loss. However it is not easy to maintain loss-free transfers for millions or billions of signs.

In the analogue form however, copying would be equivalent to a new recording, whereby the conversion of a measurement in one measurement interval is undertaken again, and this is changed again and possibly distorted. So while digital recording technology has to solve similar problems of precision as analogue technology, digital storage, unlike the analogue equivalent, does offer the possibility of almost arbitrarily increasable precision through coding and redundancies. In the case of analogue technology, precision can only be increased via the material technology, the material of recording technology and mathematical filter methods (e.g. FFT). Digital technology increases the reliability – thereby consuming resources however - of storage and copying by orders of magnitude, physical quality errors and distortions can be almost arbitrarily compensated by precision increases in the medium itself, which is impossible with the analogue equivalent. However, a completely flawless coding can never be achieved, even with digital technology (cf. Coy 2001, Pflüger 2003). Working purely in digital form is only possible in our heads (and using pencil and paper), in the symbolic arena, not however with a material computer, where the physical embodiment is paid for by a loss of precision.

3 Typography / Pictures

Number and Script (cf. Krämer 1997, Coy 2001).

Script and numbers are digital media which work with signs, that is symbols. They have different origins: Roman/Arabic numbers originated in Mesopotamia from cuneiform script differentiated by counted objects. They were later detached from the context and became abstract objects. The Babylonians and even the Romans still needed new symbols, a potentially infinite supply of signs to represent natural numbers. It was only the invention of zero in the 8th century in South China, Indochina or India which made the representation of digits possible and as a result Kaukasian/Arabian algebra (Al Kwarizmi) with a finite number of axioms for arithmetic operations. Decimal digit notation only requires a finite alphabet, even if a finite number of digits is not sufficient for numbers of arbitrary length. Figures and numbers are independent of languages, without linguistic reference and so they make up a universal sign system, are universal character set in the sense of Leibniz. It is (only) in the typographic sense that number representations in digit systems come close to words made up of letter atoms, through their use of figure atoms.

Roman letters on the other hand originated in the Mesopotamian pictographic system of writing, which became simultaneously more abstract and symbolic with syllable script and finally letter script, until the symbols had become detached from the pictures and stood for sounds, vowels or consonants. Script therefore originated, in contrast to the abstract representation of numbers, as a visualisation of spoken language: drawings that were used to create pictures of words and ideas were schematised, thereby creating symbols for words, syllables and finally letters.

Unlike pictorial scripts, and unlike numbers originally, as W. Coy (2003) points out, a finite supply of signs has always been sufficient for the alphabet. Francis Bacon standardised script alphabets and numerical digit systems through his discovery of the codeability of letters as 0 and 1. Leibniz took up this point and regarded the divine nature from then on as binarily construable and calculable, and he devised a decimal calculating machine for arithmetic operations. C. F. Gauss's telegraphic experiments with digital codes led to their electrification. Numbers were therefore readable as words and words as numbers, and the already imaginable computer did not differentiate between them (on a machine level). Simultaneously in 1936, Konrad Zuse's first electric computer based on digital technology was built and Alan Turing's concept of a universal machine which could carry out all calculable functions arose. It was only later (1954) that he saw in it the "paper machine" capable of "intelligent thought" and developed the Turing Test as proof.

Pictures

While numbers and script, in fact all symbols, are clearly syntactically and semantically definable, this is not true of visiotypes (Pörksen 1997) (pictures, tables, curves, visualisations). The question of what a picture actually is, apart from a meaningful flat, is more difficult than at first assumed, and there is no simple, widely accepted definition. The most widespread opinion – both in everyday life and in traditional philosophical picture theories – assumes that the specific element of a picture is based on either the similarity of picture and pictured object or the involvement of real objects in the creation of the picture. These similarity and causal theories, however, offer no satisfactory explanation of pictorial representation. Similarity and causal theories have grave deficits (Steinbrenner 1997). Although the similarity thesis is usually formulated correspondingly, it is certainly wrong to assume that the pictured object, e.g. a person, bears similarities to the picture, e.g. a photograph, as such. A photograph is more similar to every other photograph than to the person photographed. It is therefore not the objects themselves that are compared to one another, but the pictured object is similar to the object represented in the picture. But that is just an alteration of the difficulty: “Against this form of similarity theory, which assumes that the bundles of light rays broken in perspective which are caused in the eye by the picture, respectively the pictured object, are the same or extremely similar, speaks the practice of observing image and reality. We look at neither images nor the objects with a fixed eye” (ibid., p. 21). Causal theories of picture analysis see the reproduction relationship as explained through a cause and effect relation, i.e. the object represented in the picture is – according to this understanding – caused in a significant way by the real object. The picture must therefore be understood as a direct effect of the real object and – seen in reverse – the object is seen as a (partly) causal factor in the creation of the picture. Yet even the causal theory cannot serve as a model for defining the nature of pictures, as a causal relation can only exist between objects existing in the same time and space. This condition is not fulfilled on the part of the copy, especially the computer picture. Furthermore, not every effect of a causal relationship is necessarily a copy. The type of cause and effect relation between object and picture would have to be exactly defined for the type of copy – which is impossible, or only possible in hindsight and therefore not generally practicable.

Flusser (1983, 1992, 1995), who is primarily concerned with hypotheses for reforming our thinking through the computer and information technology, describes pictures as “significant surfaces” which mediate between the world and the human being. The world is not directly accessible for the human being – pictures serve the purpose of making the world imaginable for us. The significance of pictures lies on the surface and can be understood in just one look. The attempt to “deepen” the significance of a picture while looking at it does not involve either the person looking or the (temporary, i.e. recalled little by little) structure of the picture. When the eye wanders across the picture (Flusser calls this optical inspection “scanning”) to deepen its meaning, the interpretation of the picture achieved is made up of the picture structure and the intention of the observer. Pictures are not – as for example numbers - denotative symbol complexes, but connotative ones, i.e. they offer scope for interpretation. While the eye wanders across the picture’s surface it registers one element of the picture after another and produces meaningful relationships between these elements. In the thus created meaning complex, the elements grant one another mutual meaning. The “deepened”, “read” meaning of a picture is structured in space and time, but differently to the meaning of linear texts.

Contemplation versus Symbolic Formalisation

In the history of the sciences, contemplation had an epistemic character up until Renaissance times. The analogy between the natural and the simulating relied on contemplation. The term “contemplation” can mean both the act of cognition (visual perception) and the object of contemplation (the picture). Contemplation can be associated with clarity, with visual perception or with Kant’s concept of “pure” contemplation. Various scientific disciplines, such as phenomenology, developmental psychology and the cognitive sciences, analyse the “forms of contemplation” and thus also whether they can be formalised.

Bettina Heintz (1995) uses the term contemplation in connection with the “pictorial turn” in the sciences as the aspect of visualisation or the visual representation, whether this takes place in the head, on paper or on a monitor. The opposite term is "formalisation" or "syntactic" representation, i.e. representation exclusively on the basis of mathematical symbols and/or signs. In an epistemological respect, contemplation can take on a cognition-leading, i.e. heuristic, or a cognition-substantiating function. The formalisation in and through mathematics has increasingly suppressed contemplation as a method of substantiating cognition. It was David Hilbert who initiated the radical formalisation programme of mathematics, with his abstract understanding of mathematics in which mathematical subjects and their relationships to one another are defined purely immanently, via the axioms, and no longer relate to the world beyond mathematics. His understanding of axiomatisation means that it no longer has any real world or evident character, and its formalisation takes place in the formal language ZFC, as signs or chains of signs. It is this sign level on which formalistic mathematics operates and which has banned contemplation from mathematics. In the place of a concept of truth based on content, which would be defined by correspondence theory (Tarski semantics), a purely formal concept of truth has arisen. The process of operating with these signs, their combination and transformation, is purely mechanical in principle. From this point, it is not far to Turing's analysis of the concept of algorithm[5] as a purely mechanical deterministic process which can be made up of a few basic operations. The constructive aspect is the precise definition of the concept of algorithms in the Turing machine, which, like other formalisations of the calculable functions, forms the theoretical basis of the computer.

4 Signs and Symbols

Charles S. Peirce attempted to formulate a general definition of the elements of signs and their functions in his semiotic theory. According to this theory, a sign is described as a relationship with three components, the means of description (signifier), the object described (signified) and the interpretation of this description (by an interpretant).

The best examples of such signs are words, as descriptions for objects, interpreted by humans in a hermeneutic manner, or natural numbers as (abstract) objects, e.g. represented in decimal notation as signifiers, classifiable quantity descriptions when interpreted as counted quantities. An alternative interpretation for the same object class would be the mathematical interpretation through the representation of the semantics of natural numbers in Peano axioms.

Peirce 1894, however, differentiates between 3 types of signs, the first two of them in only two-component relations:
"There are three kinds of signs. Firstly, there are ***likenesses,* or icons**; which serve to convey ideas of the things they represent simply by imitating them. Secondly, there are ***indications,*** or indices; which show something about things, on account of their being physically connected with them. Such is a guidepost, which points down the road to be taken, or a relative pronoun, which is placed just after the name of the thing

[5] Alan Turing: "On computable numbers. With an application to the Entscheidungsproblem" (1936)

intended to be denoted, or a vocative exclamation, as "Hi! there," which acts upon the nerves of the person addressed and forces his attention. Thirdly, there are ***symbols,*** or general signs, which have become associated with their meanings by usage. Such are most words, and phrases, and speeches, and books, and libraries."[6]

The computer is able to process signals and data interpreted by humans as signs, language, images, etc.. F. Nake therefore emphasises that both aspects of the computer, that of a transformation instrument and that of the use as a medium of representation and communication, stress the sign character. He refers to the computer as a semiotic machine and to computer science as technical semiotics, with two interpretants, the human and the machine, whereby the latter behaves as a determination.

It should be pointed out that the conception of the computer as a semiotic machine is misleading, as Peirce's semiotics reflects the social dimension of the sign process through their pragmatic perspective, but computers are excluded from the semiotically mediated interplay of sociality and self-awareness by their constitutional solipsism. A decisive factor for understanding computers (and machines in general) is the index as a "degenerate", that is a merely two-sided sign relationship, without interpretants. Peirce differentiates, with regard to the object side of signs, between the 'degenerate' sign icon (similarity relationship with the signified object) and index, (causal relationship with the object signified), and the actual interpretable symbol to be interpreted (conventional relationship to the signified object). In as far as a computer causally processes signals interpretable as data, we would describe it rather as an indexical machine (if this were not a pleonasm). "An *index* is a sign which would, at once, lose the character which makes it a sign if its object were removed, but would not lose that character if there were no interpretant. Such, for instance, is a piece of mould with a bullet-hole in it as sign of a shot; for without the shot there would have been no hole; but there is a hole there, whether anybody has the sense to attribute it to a shot or not" ('Dictionary of Philosophy & Psychology' vol. 2, CP 2.304, 1902).

Threfore the interpretation of the causal connection between input and output of a computer is again up to a human interpreter. The computer does not work with models or pictures, nor with signs, but with signals without symbolic value; we humans only interpret these as signs, for numbers for example. The ability to calculate is based in turn on a physical constancy, causality, in the chip or transistor, equivalent to the basis of degenerate sign relationships. If we take a further step towards pragmatism (here for example image processing), in view of which Peirce's semiotics was designed after all, the question of the meaning of computer pragmatism arises; that is the question: can computers act (without basic sign processes, such as representation of the target as a target etc.)? It is possible to implement (consciousness-analogue) representation processes in machines, as machines can compare sections of the environment, that is external factors, with internal structures one can understand as representations, through specific feedback effects, and then intervene in the environment or modify the representation according to certain defined targets. However the defined target itself is an external guideline – out of a lack of self-awareness which would enable the target to be represented as an (own) target. The problem in regarding com-

[6] Here we can recognise the similarity and causal relationship used to define the concept of the picture.

puter science as "technical semiotics" appears to be that we can only understand the constitution of self-awareness (nowadays on both a philosophical and a neuropsychological basis) in close connection with sociality. Peirce's semiotics reflects the social dimension of the sign processes.

It therefore makes sense to differentiate between various levels and possibilities in the description of the computing method:

1. on the physical, material level, the computer is only able to store, copy, transform and transport signals, Peirce's indices. These are not interpretable by humans. However they can – with precision losses – be converted into discrete numbers, symbols, whose material representations can only be other signals, but do appear as signs on an output medium.
2. Signs, symbols, numbers and letters are, however, the media with which humans actively operate, programme and write on a computer. They may therefore interpret the computer as a sign-processing machine, although there is nothing it does less than this. Formalisation is after all the essential method of computer science for processing and symbolising real-world phenomena so that they become accessible to algorithmic treatment, i.e. can be supplied to the computer-typical signal transformation processes. For this reason, the computer science method (the method used by humans, not that used by machines: the latter is signal transport and signal transformation) is usually described as symbol processing: software is symbolic code, the connection of software and hardware overcomes the jump from sign to analogue signal from the human interpretation via discrete-analogue representation on E/A media, which has to overcome a similar jump to hardware representation.
3. An intermediate level is often introduced between interpretable symbols and signals, that of data, which concerns number values, removed from the material level (for example grey tones, or integrated values on the receptors of a tomography drum), which however can only be interpreted as purely numerical values, while the actual inherent information to be interpreted is only accessible after further processing or through integral consideration.
4. Formalisation and symbolic representation do not, however, describe computer science methods exclusively: a) the so-called "pictorial turn" from text to picture, which with increasing memory space also provokes paradigm changes within computer science, cannot be described as symbol processing, b) for the paradigms imported from biology and physics (artificial neuronal networks, gene or DNA computing, quantum computing), sign processes are no longer constitutive (even when DNA strings pair up due to "letter matches"). Connectionistic systems for example represent objects as total system states in the best case, the states of individual nodes and edge values cannot, similarly to pixel values, be sensibly interpreted other than as meaningless signal values. Meanings for humans then appear as holistic appearances (pictures) or outputs which are more complicated to interpret, for which constructed aids must be made available.

Peirce's concept of signs is also applicable to (always discretely constructed) computer pictures, however it delivers no means to differentiate between image and text. The individual signs, pixels and voxels, of which a computer picture is constructed, are degenerated in the sense that they cannot be sensibly interpreted individually other than as grey value, and so only actually stand for the object "data item in pixel form" itself. Yet within the total interplay of spatially structured data, grey values for example, they do constitute meanings that can be interpreted by humans. In this context, connotative aspects of space and time, such as the relationship of meaning between pixels/voxels and the total picture, appear more important (Flusser 1992).

We can therefore observe that the epistemological linking of the computer and semiotics (computer science as technical semiotics, computers as semiotic machines) produces several problems, loses all precision in the transition to some hybrid computer sciences such as bio-computing, and becomes irrelevant in connection with the increasing usage of the computer as a picture-generating medium.

"Arbitrary Complexity" of the Discrete[7]

The discreteness of the software medium causes various new complexity phenomena, which lie partly in the medium itself and partly in the interaction between humans and computer systems. The first class of phenomena is caused by the fact that discrete functions lack that pleasant characteristic of the continuous of being "robust", i.e. that small changes in input values can cause large changes in output values, huge uncontrolled leaps all the way to a crash. Incorrect programmes, and all "real" programmes are not correct, can therefore become chaotic. Incompleteness of specifications also causes unpredictable effects of software. Similarly chaotic behaviour is caused by the non-linearity of relationships between individual modules (i.e. the uncontrollability of side-effects on interface variables) and between program, environment and hardware parts. These are in principle characteristics of the medium, not risk potentials which could be controlled and protected against by any verification methods and test-runs, however good they were – if in fact any existed. Risk management of software and embedded systems must tackle this phenomenon in a different way than by attempting to verify the entire software. Localising the risky terminals and verifying such parts is one of the possibilities most closely located within the methods of computer science; most, however, are in the material areas of software embedding and/or organisation.

The second group of complexity phenomena mentioned result from the problems of divergence between human perception, which runs in analogue structures, and the abstractness, discreteness and the aforementioned non-robust nature of computing processes and functions. Correspondingly, programmed machines do not behave as humans expect them to. This makes it more difficult for humans to understand, that is to imagine the results ex ante, to assess the correctness of results, to assess their effects and possibilities. The consequences include the possibility of faulty operation or incorrect reactions to risky situations, whereby the problems of the initially mentioned complexity phenomena are doubled.

---

[7] Richard Brooks, Schinzel 1998

5 Epistemic Change from Text to Picture

Pictures and symbols have been cognitive carriers since the beginning of our culture. The textual science which evolved at the beginning of the modern age at first struggled against the picture-focussed ideologies before it. However this relationship between texts and pictures is a dialectical relationship, according to Flusser (1989): To the same extent as science fought the ideologies, that is attempted to explain them (away), it took on their ideas and became ideological in itself. The reason is that "texts admittedly explain pictures in order to explain them away, but pictures also illustrate texts in order to make them comprehensible" (p. 11). In this way, conceptual and imaginative thinking strengthen one other. The pictures become ever more conceptual and the texts ever more imaginative. "At present the highest conceptualism is to be found in conceptual pictures (for example in computer pictures), the greatest imagination is to be found in the scientific text" (ibid).
The science of today relies more than ever on the assumption that the logic of the world is linguistically, symbolically and mathematically comprehensible, rationally reconstructable (Wittgenstein). Yet it must tackle the growing particularisation, differentiation and complexity of its findings. Pictorial means of understanding can make abstract, complex subjects easier to understand or often understandable at all, even if this demands a loss of precision. One kind of measurement which cannot be cognitively grasped in any other way than through visual images is data which is gained in medicine through techniques such as CT, MRI, fMRI, PET, SPECT, MEG, etc.. Computer science reverses its own development: the expulsion of contemplation from mathematics through formalisation (as an opposing concept to contemplation), to which computer science owes its creation, has in reverse brought pictorial representation back into science, through its own acceleration and increase in complexity. The visualisation of dynamics and complexity is necessary to adapt to human cognitive abilities. This rediscovery of the visual dimension shows, according to Heintz (1995), the reversal of the process of formalisation. The current re-visualisation of scientific and medical-technical (and also mathematical) findings has its cause in the cognitively incomprehensible complexity of the data sets produced, which need to be returned to the realms of contemplation. These visualisation tendencies do not, however, necessarily mean a return of the contemplation of natural subjects, but a visual perception of virtual subjects, of complicatedly constructed artefacts, whose correspondence with the given, for example in the case of MRI and fMRI, is now only based on plausibility considerations and (in the given case) not on empirical evidence (Hennig 2001). Since computed tomography, the various procedures for viewing the inside of the body are no longer copying processes but picture-generating processes. That is they are not the products of electromagnetic rays on electrochemically treated surfaces as in X-ray photography[8], but of long, complexly calculated constructions and their visualisations. The new processes have led to unimagined diagnosis and research possibilities, also of a derived nature. They have hugely increased the potential of data sets, in particular complicated picture data sets, and also promoted the

[8] X-rays, which penetrate tissue and project absorption patterns, pose the problem that various area structures overlap on the 2-dimensional picture or are hidden under dense bones (white), making individual structures difficult to make out.

calculation and storage of derived data, enabled the deriving of further integrated data sets through the integration of further derived data, thereby opening up new fields of research in turn, etc. – an infinitely extendible process.

The invasion into the human body is accompanied by various problem complexes. The picture-generating processes work with huge amounts of data and carry out highly complex transformation algorithms for segmenting, smoothing, cleaning, etc., which can also misinterpret the material.

The increasing distance of the picture from the pictured object, i.e. the abstract character of such pictures produced via complicated processes, increases with every abstraction step, every derivation step and every integration step the susceptibility to mistakes, i.e. the possibility of picture artefacts which have no physiological equivalents. In paradoxical reversal of these facts, the representations and further picture-generating processes, such as cartographies of the body, which are not based on reproduction but on processing, interpreting and subsequent production, suggest an objective view of the body, and thereby standardisation. Therefore elements will always be involved which are not constitutive for the "living original", and the more complex and derived, the more such elements are involved. Which constructs are involved depends not only on channelling "technical" factors, but also on cultural, contingent factors, beginning with the usage and selection of technical means, in this case image-processing methods and visualisation techniques. The fact that they reduce complexity intensifies the danger of inadequate standardisation (Schmitz 2001) and that of inadequate representation prompting false ideas through pictorial artefacts (Schinzel 2002).

The aforementioned increase in pictorial as opposed to textual representation is an important element of epistemic changes in the sciences, made possible by information technology, but also within computer science itself. It is therefore necessary to take these paradigm changes into account when processing complex data. At first, as explained above, there is no difference on the level of material representation; we are dealing with analogue signals in every case. However one means of differentiating can be for example the place where meaning appears for humans, on the level of atomistic signs or on a holistic level. As a further consequence of the gradual replacement of the Neumann computer by parallel and evolutionary hardware systems, changes from linearity to multi-dimensional parallel or holistic processing of signals and data are occurring. The importance of algorithms is thereby reduced in favour of simulating and evolutive processing methods, for which empirical observation is gaining importance over verification, and thereby open construction with respect to outcome and meaning.

There is a strong interest in visual representation within the sciences which is expressed simultaneously with a need to defend or even save our language from visual elements. As a replacement of Rorty's "linguistic turn"[9], Mitchell also sees in the

---

[9] The history of philosophy can be seen as a series of paradigms of thinking. Richard Rorty introduced the term of "turns" and called the most recent of these changes in the history of philosophy the "linguistic turn". Put briefly, since the "linguistic turn" philosophy and humanities have assumed that society, as well as nature and its representations, are texts or discourses.

"pictorial turn" a reclamation of holistic means in rejection of semiotics. Just as language formerly did, the picture functions as a kind of model or figure for other things (and as an unsolved problem, because we still do not know what a picture is and what relation it has to language, nor what effect pictures have on observers and the world).

"Whatever the pictorial turn (...) is, it should be clear that it is not a return to naive mimesis, copy or correspondence theories of representation, or a renewed metaphysics of pictorial "presence": it is rather a postlinguistic, postsemiotic rediscovery of the picture as a complex interplay between visuality, apparatus, institutions, discourse, bodies, and figurality. It is the realisation that spectatorship (the look, the gaze, the glance, the practices of observation, surveillance, and visual pleasure) may be as deep a problem as various forms of reading (decipherment, decoding, interpretation, etc.) and that visual experience or "visual literacy" might not be fully explicable on the model of textuality. Decisively, however, the pictorial turn is the realisation that, although the problem of pictorial representation has always existed, it now puts us under unavoidable pressure with unprecedented strength, and on all levels of culture, from the most refined philosophical speculations to the most vulgar products of the mass media" (Mitchell 1997, p. 18f.).

Digital Pictures versus "Analogue" Pictures

The difference between old pictures, such as paintings, and new pictures such as television or computer pictures, seems to be the guiding intention in the production of the pictures. Conventional pictures were produced with the intention of making a circumstance visible. The new pictures, according to Flusser, are intended to make dots floating in nothing visible, whereby they however pretend to mean a circumstance. The essential element of all new, technical pictures is the dot structure on which they are based, their quantum mosaic character. In this sense they are immaterial. Their intention is to create an illusion.

"The new pictures (...) attempt to deceive the observer, in two ways. Firstly, they hide the fact that they are computations of dots and pretend to have the same meaning as conventional pictures; and secondly, they apparently admit to their origin on a higher level of deception, but only in order to present themselves as 'better' pictures, by pretending not to signify a circumstance symbolically, as conventional pictures do, but 'objectively', dot for dot." …"but the technical pictures do signify those clear and distinct concepts which are equivalent to the dots of which they are constructed. The technical pictures 'point a finger' at the program in the machine that created them, and not at the world out there. They are images of concepts, surfaces devoted to calculating thinking and not the concrete world. They are abstracted surfaces lifted out of abstraction in the direction of the concrete. The gesture creating them runs in a direction opposite to traditional pictures. In this sense, the technical pictures are anti-pictures" (Flusser 1995 p. 48f., 50).

Following Flusser, technical pictures therefore do not signify the concrete world surrounding us or parts of it, but are related to a conceptual universe. They are exact and "true" pictures – only they are true to their programs, not their apparently pictured object. They only signify this object in as far as the concepts involved in the program signify this object, and these concepts are already a very extensive abstraction from the concrete object. The uncanny thing about the effect of synthetic pictures is that – although the path of their creation leads across the last, dimensionless abstraction, conceptual registration, put into the form of a digital code – the picture, the model, that represents a projection from this abstraction can then be experienced in a concrete manner. These pictures refer to concrete things or phenomena only by way of the diversion via concepts which programme them. They are therefore not abstract pictures but the attempt at a concretisation of abstractions.

"The technical picture is a picture created by machines. As machines on their part are products of applied scientific texts, technical pictures are the indirect results of scientific texts", according to Flusser (1983). "This gives them, historically and ontologically, a different standing to traditional pictures. (…) Ontologically, traditional pictures signify phenomena, while technical pictures signify concepts" (p. 13).

The code in which the computer works, "thinks", is structurally very simple and functionally extremely complex. These codes allow a new, no longer linguistic form of thinking and will replace the alphabet as carrier of thinking in the foreseeable future. The code of the computer at first appears – due to its simple structure – to be equivalent to a reduced, purely calculatory style of thinking. Yet it was found that the new codes allowed computers to compute lines, surfaces, spaces and moving bodies. In this way it is possible to synthesise unexperienced objects and thereby make them experienceable. The complexity and the power of the new form of thinking lies in this designing capability.

"The (...) characteristic of the computer to be emphasised in this context is the fact that it allows not only numbers to be computed, but also these numbers to be synthesised into forms. That is a shocking invention or discovery, if we consider that calculatory thinking has penetrated deeply into the phenomena and that the latter are disintegrating into particles due to this advance. The world has thereby taken on the structure of the number universe, which poses confusing cognitive problems, if computers show that calculatory thinking can not only disintegrate the world into particles (analyse), but also put it back together again (synthesise)" (Flusser 1995, p. 280).

Technical pictures are difficult to decode. They give the impression – although this is misleading – that they do not need to be decoded at all. Their meaning seems to be automatic and therefore obvious: an automatic copy similar to a finger print, for which the meaning (finger) is the cause and the picture (print) is the result. The signified world of technical pictures appears to be the cause of technical pictures; both appear to be on the same level of reality. Technical pictures appear to be not a symbol but a symptom of the world. Technical pictures are seen - due to their perceived non-symbolic, objective character – less as pictures than as windows to the world. This leads to a lack of criticism of technical pictures, which becomes dangerous when they are in the process of replacing texts. Technical pictures are dangerous because their objectivity is an illusion (Flusser 1983).

Once Again: Computer Science = Technical Semiotics?

Exploration, representation and integration of large complex data sets make use of simulation and visualisation alongside distributed databases (Keim 2002). To remain in the context of the neuro-sciences and life-sciences, examples could be scattered data, statistical data from image analyses, parameterised data from simulations, or volume data records. This data does have formative power, but carries no meaning. In the case of a voxel with parameter values or scattered data from MRI measurements, these are not symbolic representations; they stand for no other meanings than themselves. Nor can current neuron values or propagated edge values in neuronal networks be judged as meaningful signs (for humans); the analogy to signals or impulses is more obvious in this case. It is also disputable whether the base pairs of gene sequences in DNA computing transport semantics interpretable by humans, other than the formative meaning of the biological starting material which they carry with them. My thesis is that image-processing hardware, evolutionary computers and protein chips are less semiotic machines than pattern-arranging machines; the symbolic aspect of their input/output and inner structure processing takes a back seat to signal transmission or molecule manipulation for example. Their operations only become meaningful and form-giving in concert with their holistic total effect.

Formalisation and thereby logic lose significance in contexts where pattern and pictures represent the input and output. In the same way, the relevance of algorithms as a problem-solving method decreases and if they are needed at all, then it is either as an initialisation or a processing method. They then have fewer explicit characteristics than in symbol processing. Then, the propagation of local operations and characteristics onto the global situation for example is important, characteristics can often only be proved empirically and verification is hardly possible at all[10]. In DNA, protein and quantum computing, as in many forms of connectionism, programming is reduced to initialisation, i.e. to a declarative aspect, while the (usually no longer uniform in the sense of calculability theory) computer itself takes charge of the process. In image processing, the relevance of algorithms lies between local explication and global effects, which are usually observed as physical systems and thereby construct a stubborn anti-reality. As a problem-solving method however, algorithms are on an analogue path with distributed and internet algorithms, etc. It is therefore necessary to describe the epistemological basis of computer science in a more differentiated manner than by ascribing it the quality of technical semiotics, and to differentiate the sign processes occurring according to human and technical levels of interpretation. Picture generation and interpretation are just as useful means of cognition for this task as the analysis of non-classical algorithms, computer structures and materials.

[10] Of course, these observations are far from new. Since the beginning of computer technology, images have been processed and neuronal models have been used (e.g. Perzeptron in the 1940s), and self-reproducing machines have been modelled. For example, the significance of evolutionary computer models in the context of differentiating between symbolic and subsymbolic AI (and also the adequacy of models for cognition) has long been extensively discussed (cf. e.g. Becker, 1992). Nevertheless, these epistemes have gained practical significance and with the convergence of biology and computer science also a new level of relevance, which has an effect on the scientific paradigms of computing.

References

Andersen, Peter Bøgh: A theory of computer semiotics. Semiotic approaches to construction and assessment of computer systems. Cambridge: Cambridge University Press 1990;
Andersen, Peter Bøgh: http://www.cs.auc.dk/%7Epba
Wolfgang Coy: Analog/Digital - Bild, Schrift & Zahl als Basismedien, in H.Schanze u.a.: *Bildschirmmedien*, (2001)
Flusser, Vilém: Für eine Philosophie der Fotografie, Göttingen 1983;
Flusser, Vilém: Die Schrift. Hat Schreiben Zukunft? Frankfurt a.M. 1992.
Flusser, Vilém: Lob der Oberflächlichkeit. Für eine Phänomenologie der Medien, 2., durchgeseh. Aufl., Mannheim 1995.
Grube, Gernot (1995): Modellierung in der Informatik. In: Fischer, M.; Grube, G.; Reisin, F.-M. (Hg.): Abbild oder Konstruktion – Modellierungsperspektiven in der Informatik KIT Report 125. TU Berlin
Heintz, Bettina: Zeichen, die Bilder schaffen; in: Johanna Hofbauer, Gerald Prabitz, Josef Wallmannsberger (Hrg.): Bilder, Symbole, Metaphern. Visualisierung und Informierung in der Moderne, Wien 1995, pp. 47-81
Hennig, Jürgen: Chancen und Probleme bildgebender Verfahren für die Neurologie; in Schinzel (ed.): Interdisziplinäre Informatik: Neue Möglichkeiten und Probleme für die Darstellung komplexer Strukturen am Beispiel neurobiologischen Wissens; Freiburger Universitätsblätter, 3, 2001, Rombach, Freiburg.
Keim, D.: Visual Exploration of Large Data Sets; in Visualizing Everything, Comm. ACM, Aug. 2002, Vol. 44, 8.
Krämer, Sybille (1988): Symbolische Maschinen: die Idee der Formalisierung in geschichtlichem Abriß. Wiss. Buchges. Darmstadt
Mitchell, W.J.T.: Der Pictorial Turn, in: Kravagna, Christian (Hrsg.): Privileg Blick. Kritik der visuellen Kultur, Berlin 1997, pp. 15-40
Nake, Frieder: Von der Interaktion. Über den instrumentalen und den medialen Charakter des Computers. In: Nake, Frieder (ed.): Die erträgliche Leichtigkeit der Zeichen. Ästhetik Semiotik Informatik. Agis Verlag Baden-Baden 1993, pp. 165-189.
Peirce:http://www.helsinki.fi/science/commens/die,
http://santana.uni-muenster.de/Linguistik/user/steiner/semindex/peirce.html,
http://www.iupui.edu/%7Epeirce/web/ep/ep2/ep2book/ch02/ep2ch2.htm
Pflüger, Jörg-Martin: Informatik auf der Mauer, Informatik Spektrum 17:6; 1994.
Pflüger, Jörg-Martin: Vom Umschlag der Quantität in Qualität – 9,499… Thesen zum Verhältnis zwischen Analogem und digitalem; in M. Warnke (Hrsg.): Computer als Medium „Hyperkult 12“, analog digital - Kunst und Wissenschaft zwischen Messen und Zählen, 24-26 July 2003.
Pörksen, U.: Weltmarkt der Bilder. Eine Philosophie der Visiotype; Klett-Cotta, Stuttgart, 1997.
Schaub, Martin (1992): Künstliche und natürliche Sprache; OLMS Philosophische Texte und Studien; Hildesheim, Zürich, New York
Schinzel, B.: Körperbilder in der Biomedizin; in Frei Gerlach, F., et al (ed.): Körperkonzepte; Münster/New York/München/Berlin: Waxmann 2003.
Schinzel, B.: Informatik im Kontext der Genderforschung in Technik und Naturwissenschaft; FIFF-Kommunikation 4, December 2001, pp. 19-28.
Schinzel, B. (ed.): Interdisziplinäre Informatik: Neue Möglichkeiten und Probleme für die Darstellung und Integration komplexer Strukturen in verschiedenen Feldern der Neurologie; Freiburger Universitätsblätter Heft 149, 3. Heft, 2001; 180 p; Rombach, Freiburg.
Schinzel, B. (1998): Women's Ways of Tackling the Specification Problem, AISB Quarterly (Journal of the Society of Artificial Intelligence and Simulation of Behaviour), No 100, Summer 1998, pp. 18-23.

Schinzel, B. (1998): Komplexität als Ursache für Fehler in und Risiken mit Software, FIfF-Kommunikation, 1, pp. 18-21.
Schmitz, Sigrid: Informationssysteme zu neurobiologischem Wissen – Chancen und Grenzen; in Schinzel (ed.): Interdisziplinäre Informatik: Neue Möglichkeiten und Probleme für die Darstellung komplexer Strukturen am Beispiel neurobiologischen Wissens; Freiburger Universitätsblätter, 3, 2001, Rombach, Freiburg.
Steinbrenner, Jacob; Winko, Ulrich: Die Philosophie der Bilder, in: Dies.: Bilder in der Philosophie & in anderen Künsten & Wissenschaften, Paderborn/München/Wien/Zürich 1997, pp. 13-40.
Tarski, Alfred (1983): Der Wahrheitsbegriff in den formalisierten Sprachen; in Berka, K., Kreiser L: Logik-Texte Berlin, p. 443.

Towards a Theory of Information

Wolfgang Lenski

Department of Computer Science
University of Kaiserslautern
P.O. Box 3049
D-67653 Kaiserslautern
lenski@informatik.uni-kl.de

Abstract. A solid foundation is a prerequisite for a scientific discipline to provide a modelling framework for the representation purposes in other areas. In this paper we discuss the suitabiliy of logic for this kind of application and investigate the necessary steps for the concept of 'information' to provide a comparably justified foundation for the modelling of 'real world' phenomena. This provides an analysis of the conceptual prerequisites that are required for a well-founded theory of information.

1 Introduction

Scientific development in its most general aspects is characterized by separating a discipline from a more general one be it by the evolvement of new perspectives on a field or topic, by new paradigmatic methodologies to analyze phenomena, or evolvements of other kinds that are able to through a new light on previously accepted perspectives, ontologies, or methodologies or isolate aspects for which special treatment is enabled, promised, or envisaged. In this sense most scientific disciplines have separated some time from philosophy which once constituted the unifying approach to understand (parts of) the universe. This process necessarily raises new concepts determining the evolving field or at least initiates a modified understanding of the central ones. An indispensable demand in this process is to provide a clear and solid foundation of the core concepts of the new field. This was especially in full methodological strength the challenge in logic in the 19th century which led to the problem of self-assurance of logic as a scientific discipline of its own right based on new insights.[1]

In general, a solid foundation is a prerequisite for a scientific discipline to provide a modeling framework for the representation purposes in other areas. In this paper we discuss the suitabiliy of logic for this kind of application and investigate the necessary steps that have to be undertaken for the fundamental concept of 'information' to provide a comparably justified foundation for the modeling of 'real world' phenomena.

[1] An interesting side-aspect of this problem demonstrates that bibliographies may play a major role in such situations. One motivation for the compilation of the bibliography of *symbolic* logic by Alonzo Church [24] was certainly to constitute and determine the field of logic as a scientific discipline in an extensional sense thus complementing the discipline-immanent ideas.

W. Lenski (Ed.): Logic versus Approximation, LNCS 3075, pp. 77–105, 2004.

1.1 From Philosophy to Logic

Whereas Kant still considered logic as being finally completed and fully characterized by Aristotle's theory of syllogisms (see [47, B VIII]), there have been tendencies in the nineteenth century to provide more solid grounds than philosophical considerations on 'obvious and plausible insight' into the nature of concept, proposition and consequence. In performing its 'mathematical turn' logic has finally separated from philosophy: a mathematization of logic should guarantee the wanted justification instead of legitimations based on plausibel and intelligible arguments. In this spirit the early works of Boole [14], and Schröder [84] — just to mentioned two of the most prominent ones — aimed at a construction of logic as a mathematical (sub-)theory in the spirit of current mathematics. Although Peirce was most influential not only on Schröder's later works (cf.,e.g., [73]), it was the pioneering work of Frege's *Begriffsschrift* [36] in 1879 that readjusted the overall direction of logical research. Instead of introducing logic as a mathematical theory among others, Frege attempted to constitute logic as a theory of *truth* and *proof*. It should last up to the thirties in the twenties century that a foundation of the concept of 'truth' has been presented by Tarski [88] and a clarification of the reach of the concept of 'proof' within this framework has been provided by Gödel [39]. Truth has not been invented by Tarski, though, and there is a long tradition of actual usage of this concept even in mathematics. To identify the necessary reductions of the conceptualization compared to an everyday understanding and to bring it into a shape suitable for formal (or mathematical) treatment is an absolutely outstanding achievement that can hardly be overestimated.

Concerning its suitability for modeling purposes one has to concede that logic has provided a clear foundation for *mathematics*. Thus logic has successfully provided the means for the representation requirements of mathematics. It well answers question and provides means for gaining insight into the structural properties of *mathematical* models. This works perfectly well. So naturally the question arises why it works so well for mathematics but has not proven to provide valuable instruments to model real world problems as well. One quick answer is given by the fact that it relies on *static* principles: mathematical truth has always been considered as being *necessarily* true, i.e. in the aristotelean sense true *at all times*. In systems which an inherently dynamic character, however, the purely logical approach has not proven to be comparatively successful. Most real world problems belong to this category.

There have been attempts to apply the logical methodology to this kind of problems as well. Some approaches in the philosophy of science. e.g., focus on *axiomatizable* theories (cf., e.g., [87] or [69]). But this line of scientific investigations have not resulted in a widely accepted re-adjustment of research aims and methodologies. Another reaction is fuzzy logic introduced by Zadeh [94]. It has led to successful applications especially in control theory but has also not proven to provide a general methodology for the modeling of 'real world' phenomena. The pure logic-based approach seems too restricted for such purposes and computer science has essentially taken over this kind of challenge.

1.2 From Logic to Computer Science

Along with the scientific and technical development in the middle of the last century *computer science* has evolved which complemented theoretical mathematical considerations going back (at least) as early as to Leibniz with essential contributions of Dedekind 1888 [25] (who proved that — what is now called — the scheme of primitive recursion does define a unique function), Kleene, Gödel, Church, and especially Alan Turing [91] by technological progress under specific influence of von Neumann.

Its scope has been broadend since then dramatically and the subsequent development has brought computer science into a major position for providing modelling tools to capture 'real world' phenomena. The most noticable consequence, however, that has influenced society dramatically is centered around information technology. There are discussions of the chances of getting informed in ways nobody had dreamed of some few decades ago and critical voices are mentioning the danger of information overload which leads to an increasing demand for specifically focussed information.

Whereas a clear foundation of the *mathematical* parts of computer science via 'truth' and 'proof' are implicitly provided by the logical roots of mathematics as a result of the 'Grundlagenkrise' in the 19th century, a comparable foundation of the basic concept of 'information' for computer and information science has not been shown so far.

An immediate approach to model 'information' on the basis of the (unrestricted) logical framework leads to undesirable results. A counter-argument against an attempt to rely on the full power of logic is especially provided by Hintikka's concept of omniscience (see [41]) where knowers or believers are *logically omniscient* if they know or believe all of the consequences of their knowledge or beliefs. Accordingly, knowing the axioms of (some extension of) ZF-set theory, e.g., should entail knowing of all consequences thereof including knowledge of inconsistencies.

Compared to logic we are thus currently rather in a 'pre-Tarskian' stage of development for the concept of information. As we have noted beforehand also the concept of 'truth' has be subject to reductions of the full scope of its understanding. In the following we will investigate perspectives for conceptualizations of 'information' that could pave the ground for a *theory* of information on which extended instruments for structural modelling purposes of 'real world' phenomena could finally be based on in a most justified way.

2 The Context of the Concept of Information

Even a short preoccupation with possible meanings of the term 'information' will immediately bring to light that very different and wide-spread understandings of the concept of information have evolved. "There is no accepted science of information" so the opening remark in the preface of the monography of Barwise and Seligman [7] — not to speak of a broadly accepted *theory* of information.

Hence the indispensable first step towards a theory of information is to clarify the concept of information in its different shades. Accordingly, we start with a survey of conceptualizations of 'information', exhibit essential aspects of and finally systematize insights into the nature of information that have shown in the literature. This is, however, not meant as a comprehensive study of the concept of information in the different contexts; such an attempt would certainly deserve a more detailed analysis.[2] Instead, we trace understandings and usages of the terms with to aim to determine those constitutive properties that have to be taken into consideration for a prospective theory of information which has certainly to be developped in conformity with the (underlying principles of the) common usage of the concept as discussed in the relevant communities.

In this sense this chapter is intended as a guide providing orientation in the landscape of conceptualizations of 'information', putting landmarks, and identifying points of view. A coherent interpretation of all these so far isolated considerations according to a philosophical background theory is the very prerequisite for a subsequent well-founded characterization of the concept of information.

To begin with we have to acknowledge that a conceptualization is always part of a view of the 'world'. As such, it cannot be understood without the context in which it is developped and to which it contributes. Accordingly, this section also exhibits key-concepts that are closely related to 'information' and thus must be included into a theory of information as well.

2.1 Concepts Necessarily Associated with Information

To begin with we record that the concept of 'information' is closely tied to the concepts of 'knowledge' and 'data'. This seems to be a shared understanding in all contexts dealing with conceptualizations of 'information' which is reflected in practically all publications on the topic. The concept of information being tied to data is apparent already in the early publications that focus on the relation between computer science and society, their mutual influences and general perspectives. An early definition which sounds typical for the understanding of the relation is, e.g., given in 1973 where Sanders [81, p. 6] already refers to a allegedly general understanding when summarizing:

> *information* is generally considered to designate data arranged in ordered and useful form. Thus, information will usually be thought of as relevant knowledge, produced as output of processing operations, and acquired to provide insight in order to (1) achieve specific purposes or (2) enhance understanding.

In this definition information is understood as the result of a transformation process that relies on general understanding of terms like *useful*, *purpose*, and *understanding*, etc.. Interestingly enough, information is then further described

[2] For a more detailed exposition of the concept of information in the context of information and library sciences and information retrieval the gentle reader may, e.g., consult [44].

as the result of a process that "reduces uncertainty" [81, p. 6]. This additional remark hints at the first line of understandings of 'information' which has been worked out by Shannon and others (see below) and must be considered as an attempt to turn the definition into a formally treatable form — at the cost of some overly simplified conceptual reduction as we will see in the following.

The connexion between 'information' and 'data' then remains generally acknowledged up to now. Evidence for this is given by the fact that a clarification of these concepts along with their relationships is especially demanded by the joint ACT and IEEE *Computing curricula 2001 Computer Science* [1] where

> compare and contrast information with data and knowledge

is explicitly noted as a learning objective in IT courses [1, p. 134]. Precise definitions of the concepts, however, are not given in the curricula as it is intended to constitute a list of indispensable topics to be covered by respective courses. So this remark will provide us with the guideline to the central concepts that must be clarified for a formal theory of information as well. But beforehand we look for some more candidates that might also be intimately connected with the concept of 'information'.

It is not generally agreed, though, that 'data', 'knowledge', and 'information' constitute an exhausted list of the fundamental concepts that are necessarily tied to each other. There are mainly two suggested extensions of this enumeration: *wisdom* and *practice.*

Practice. To fill the framework of the ACT and IEEE *Computing curricula 2001 Computer Science* Denning [27] tries to define the concepts 'data', 'knowledge', and 'information' from the viewpoint of curricula for information technology education for IT professionals.

> "Information" is the judgement [...] that given data resolve questions. In other words, information is the meaning someone assigns to data. Information thus exists in the eyes of the beholder. [27, p. 20]

In the course of his exposition Denning explicitly extends the triade of 'data', 'knowledge', and 'information' by a fourth dimension named 'practice' denoting 'embodied knowledge': this describes patterns of action that effectively accomplish certain objectives with little or no thought [27, p. 20]. It should be noted that this fourth dimension seems not to be of comparable epistemic status but instead captures some abilities of the human mind to disburden attention from well-established actions (see [38, cf. esp. volume 3.1, p. 35]).

Wisdom. Another suggestion for a concept that should be related to information originates to Ackoff. In his influential approach [2] the relation between data, knowledge, and information is understood as a conceptual transition from data to knowledge and from knowledge to information initiated by an each time increased understanding. Moreover, this transition admits one more step, namely from knowledge to wisdom: understanding relations leads from data to information; understanding patterns results in knowledge; even understanding principles gives raise to wisdom.

It should be noted that in this approach the concept of information is prior to the concept of knowledge: knowledge is created out of information as opposed to [81] where information had been characterized via knowledge.

It already appears at this stage that a closer interrelation between data, knowledge, and information is indeed fundamental whereas an *epistemic* dependency between concepts like practice or wisdom on one hand and data, knowledge, and information on the other hand remains at least doubtful. Beyond the mere conviction that data, knowledge, and information are closely related to each other there is, however, no commonly shared understanding of the *kind* of this relationship. It is our intention to contribute to this discussion as well.

3 Philosophical Background

So far we have exhibited the core concepts for a theory of information. This section now is devoted to philosophical investigations of the epistemic status of these concepts along with their relationships. The basic situation is comparable to the situation in logic concerning the concept of truth where Tarski [88] has explicitly referred to foundational philosophical positions, namely the semiotic approach as the philosophical ground for his pioneering theory of truth. This means that he explicitly wanted to prevent his approach from being some kind of ad hoc or based on a specifically taylored background. In this sense a philosophical reflection seems inevitable and indispensable at the same time.

Tarski has then explicitly chosen a *semantic* foundation of the concept of truth. This seems to be a canonical decision, and it is doubtful whether he actually would have had another choice. In the following we will investigate the epistemic status of the concept of *information*.

Our approach is committed to the semiotical theory in the tradition of the pragmatic philosophy after C.S. Peirce [74] as well.Pragmatism (or pragmaticism as it was called later on) tries to react on deficiencies identified in the philosophical tradition where especially the gap between pure epistemology and practice could not be bridged in a philosophically sufficient way.

While even a rough outline of the philosophical insights would be out of the scope of this work, we will instead just sketch some basic principles of this approach that may reveal the influences of this position to the theory of information we are developing in the sequel. In this spirit it is just intended to introduce the main concepts and to evoke the central ideas of the position laying behind. This is definitely not to be understood as a careful philosophical analysis but rather as a brief review of the basics of Peirce's philosophy. Readers who are not really acquinted with the fundamentals of this approach and/or are more interested in the topic should rather consult some introductory textbooks such as for example [43], [72], [33], [26], [67], [37].

3.1 Pragmatism

According to the philosophical tradition originating from the (idealistic) position of Kant, knowledge is composed via perception and subsumption of concepts under categories. While sharing this general approach (at least in principle) so far, Peirce emphasizes at this point that the actual establishment of knowledge is performed by individuals and results in an orientation in the 'world'. This is seen as a process in which first concepts are formed and finally knowledge is established. These concepts are closely tied to indended possible actions by that very person. This idea is expressed by the *pragmatic maxim*:

> Consider what effects, that might conceivably have practical bearings, we conceive the object of our conception to have. Then, our conception of these effects is the whole of our conception of the object. [74, paragraph 5.402]

According to the pragmatic maxim, concepts are thus constituted through the effects they may evoke. The conceived effects then form the basis for our conception of objects which gives raise to knowledge. It remains to show in which way or, more precisely, on what basis this is performed.

Now what initiates our awareness about phenomena are *signs*. Signs are considered as the very basic constituents of all epistemic processes. Hence knowledge is especially represented by signs. As a consequence, this view essentially demands a theory of signs along with a theory of understanding and interpreting these which is developped in the field of semiotics.

The interpretation of epistemic processes as part of a general theory of signs on the other hand necessarily requires a fundament to base epistemic considerations on. According to the classical philosophical tradition especially in the spirit of Kant's epistemology, this requires the detection of a system of *categories* that determine the fundamental aspects of any possible dealing with signs. Peirce introduces three universal categories i.e. categories that are always present in every epistemic process. They are described in his third lecture on pragmatism 1903 as follows [74, Vol. 8, paragraph 328; A Letter to Lady Welby, 1904]:

1. *Firstness* is the mode of being of that which is such as it is, positively and without reference to anything else.
2. *Secondness* is the mode of being of that which is such as it is, with respect to a second but regardless of any third.
3. *Thirdness* is the mode of being of that which is such as it is, in bringing a second and third into relation to each other.

Understanding may now be characterized as a principally uncompleteable process of the effect of signs for an interpretant. This process is called *semiosis* (cf. [74, paragraph 5.484]).

> [Semiosis is] "an action, or influence, which is, or which involves, a cooperation of three subjects, such as a sign, its object, and its interpretant, this tri-relative influence not being in any way resolvable into actions between pairs." [74, paragraph 5.484]

It basically implies a triadic understanding of signs as claimed by the categories. Because of their status as universal categories, they are basic for all epistemic process. Especially, the theory of signs must reflect this. Hence the theory of signs as sketched above is also subject to a triadic relation. This has been pointed out by [19, 99]:

> A sign, or Representamen, is a First which stands in such a genuine triadic relation to a Second, called its Object, as to be capable of determining a Third, called its Interpretant, to assume the same triadic relation to its Object in which it stands itself to the same object. The triadic relation is genuine, that is its three members are bound together by it in a way that does not consist in any complexus of dyadic relations

It is important that there is not only one stage of triadic interpretation of signs. The process of a 'semiosis' is a multi-stage process in principle: a sign in its triadic specification may well be an object of consideration with a specific viewpoint in itself. This turns one triadic sign into a representamen for another process which is another triadic sign with another object and another representant. It is worth mentioning at this point that the *interpretand* in Peirce's theory of signs is the *sign* created in the mind of a person and not the person itself.

In general his categories along with his semiotical analysis are meant to provide the means to bridge the gap that Peirce identified in the Kantian epistemology between concept on one hand and their relation to (possible) actions on the other hand. It is the process of the semiosis that should overcome this fundamental restriction detected in the philosophical tradition.

According to the pragmatic semiotics we have to deal with signs, concepts, and objects together with their relations. Concepts on the other hand must not be separated from the process of establishing knowledge and according to the pragmatic maxim [74] thus may not be thought without possible consequences or actions associated with them in our imagination.

Peirce's theory of signs may in short be explained via the following statement in [19, p. 99]:

> A sign, or representamen, is something which stands to somebody for something in some respect or capacity. It addresses somebody, that is, creates in the mind of that person an equivalent sign, or perhaps a more developed sign. That sign which it creates I call the interpretant of the first sign. The sign stands for something, its object. It stands for that object, not in all respects, but in reference to a sort of idea, which I have sometimes calles the ground of the representamen.

Philosophical ideas pre-structure the domain of things that may be dealt with. So these structures implicitly underly all phenomena and thus have to be taken into consideration in order to establish a cognitively adequate (information) model. But since the main focus of this paper is not on problems of epistemology, we will interrupt the discussion of the philosophical background at this point. It is rather our concern to base our theory on clear philosophical insights and especially to counter the kinds of reproaches as for example stated in [56, p. 61]

> Diese [...] Orientierung hat die Informatik vor dem Hintergrund einer vielerorts fehlenden Bereitschaft zur Auseinandersetzung mit ihren methodischen Grundlagen davon abgehalten, sich den Problemen der Praxis in einer methodisch tragfähigen Weise zuzuwenden.

In this context the present work is especially meant to face this reproach and to develop a methodologically well-justified theory of information instead to contribute to a philosophical discussion. To do so we have to bridge the gap between the philosophical considerations and conceptualizations that can actually be implemented into a real information system.

The investigation of consequences that may serve as a concrete basis for our intended application requires a transformation of principles of rather epistemic nature into conceptualizations that admit a more practical treatment. Such an interpretation will be presented in the following section.

3.2 Morris' Analytical Reductions

Whereas Peirce's categories had been meant as universal philosophical categories underlying and guiding every epistemic process, in the succession a purely analytical reinterpretation of the semiotical relationships on the basis of the three categories by Morris (cf. [66]) has been worked out. In the following we will adopt this analytical reduction for our considerations. Accordingly, semiotics as the general theory of signs has three subdivisions [65] (see also [20]):

(1) *syntax*, the study of Òthe formal relations of signs to one anotherÓ,
(2) *semantics*, the study of Òthe relations of signs to the objects to which the signs are applicableÓ,
(3) *pragmatics*, the study of Òthe relation of signs to interpretersÓ .

In this analysis some possible relationships are missing. One could certainly think of the relation between an interpreter of a sign and the objects to which the signs are applicable ([79]). This leaves room for further investigations based on these dimensions.

3.3 A Semiotic View of Information

In this section we discuss the semiotic nature not only of information but also of the related concepts of knowledge and data. We will show that the semiotical approach provides a theoretical basis for an unified view on data, knowledge, and information. The common basis of these concepts is provided by the abstract concept of a *sign* which is considered as an ontological unity.

Now signs in an epistemic context are subject to analytical considerations according to Morris' semiotical dimensions. This results in the following conceptual coordination:

Data. *Data* denotes the *syntactical* dimension of a sign.

As Morris' semiotical dimensions are analytic abstractions that are derived from Peirce's categories, the universality of the latter as a whole cannot be circumvent and is always present in all mental activities. So isolating just one dimension of a comprehensive semiotic analysis can only be done artificially — and only for analytical purposes. One may certainly focus on projections of the whole to one of its dimensions, but these are then just restricting perspectives neglecting the other dimension that are nevertheless associated with the object as well.

In this sense *data* denotes the organized arrangement of signs with emphasize given on the structural or grammatical aspect only. Moreover, just arbitrary arrangements of signs are not considered as data. This implies that data are always derived from an understanding of a part of the world, not withstanding the fact that the result is basically a sequence of pure signs.

In view of the pragmatic maxim (see page 83) this implies that the concept of data must be understood with respect to its actionable intentions — its usage. According to their syntactic dimension data are not interpreted, but interpretable. The adherent interpretation, however, is external to the syntactic structure; data provide the raw material for subsequent interpretation.

This aspect implies several things. At first, the creator must make sure that a subsequent interpretation is feasible. This is performed by using an organizational structure, a grammar, that refers to an understanding which can supposed to be taken for granted (at least to some extent) in the community to which the intended contents is addressed.

The (re-)construction of the underlying ideas requires at first the usage of a commonly understood organizational form, a grammar. Such a grammar is then endowed with a shared interpretation within a community. So it is this procedure of interpretation which is made possible by referring to a organizational form that in a community is associated with a form of interpretation.

However, it is the usage of a commonly understood organizational form, a grammar, which is *external* to the data that allows this (re-)construction.

Knowledge. *Knowledge* denotes the *semantical* dimension of a sign.

Knowledge is generally understood as being certain.[3] This certainty may be true (cf., e.g., [5, p. 34]), or at least justified in some sense (see [7].

As for syntax, the universality of Peirce's categories also leads to consequences for their analytical abstraction in semantics which abstracts from per-

[3] To relate knowledge to certainty goes (at least) back to Descartes [28]. Descartes in his quest for solid grounds for certainty of reasoning essentially relied on mathematics as providing a convincing model. A similar understanding has also been taken over by Kant when writing

> Endlich heißt das sowohl subjektiv als objektiv zureichende Fürwahrhalten das Wissen. ([47, B 850/A 822])

and has subsequently been broadly acknowledged; see for example [32] or [49].

sons that actually ascribe its contents. It is the very process of analyzing this abstraction process that is subject to the category of *thirdness* as well. In this context it amounts to analyzing the grounds on which this abstraction is performed — and especially by what reasons a conviction (necessarily) be shared by a community which results in a common acknowledgement of the abstraction process (and hence of its results as well) establishing (at least some sort of) intersubjective commitment. This investigation is necessarily part of the procedure to *account for* semantical relationships.

In this sense we may analyze characteristics of the *kind of abstraction* involved in the actual generation of and subsequent attribution as knowledge. Accordingly, the abstraction leading to various degrees of certainty may be

- universally valid
  This would be ideal for a pure semantic characterization and is in general the pretension of *truth*. However, even Tarski's semantic theory of truth [88] in mathematics has not experienced general acceptance as a universal methodology and remains bound to Hilbert-Tarski-style of mathematics; see, e.g., constructive mathematics in the sense of Brouwer. As there are no other grounds for a universally accepted methodology currently being visible, this claim must be considered as a 'regulative idea' in the sense of Kant that may only be approximated in reality.
- depending on presuppositions
  This hints at a community sharing the presuppositions. As a consequence the abstraction determinuing the semantic nature of the knowledge is only acknowledged by this community.
- only personal.

This gives raise to several dimensions or degrees of justification that may be distinguished. Whereas 'true' and in a similar sense 'valid' or 'justified' comprise a universal claim of justification, there may exist personal knowledge.

Information. *Information* denotes the *pragmatical* dimension of a sign.

Being of pragmatical dimension, information demands an interpretand to perform an interpretation. Information is bound to a (cognitive) system to process the possible contributions provided by the sign (the data) for a possible action. Such an action on the other hand is by no means arbitrary but 'purposeful', i.e. driven by an intention. It is this very aspect that is vaguely described as the "biased" nature of information (see, e.g, [55, p. 265]). It has become usual to denote such system by *agent* be it a human or not. This is described in [29] as follows:

> it is common among cognitive scientists to regard information as a creation of the mind, as something we conscious agents assign to, or impose on, otherwise meaningless events. Information, like beauty, is in the mind of the beholder.

Especially, information inherits the same interpretation relation as knowledge with the difference that the latter abstract from any reference to the actual performance of the interpretation. According to the universality of Peirce's categories this implies that the relation to 'possible actions' is just neglected in the concept of knowledge. But then the question raises what kind of (abstracted) relations may actually result from such an abstraction: it must result in some *generalized* experience which in turn brings up the question of the kind of justification of such an abstraction.

This is the epistemic position which shows that knowledge and information are indeed closely tied together. The systematic question whether knowledge or information is prior then amounts to the question of the relevance of the actual performance of the interpretation:

- For knowledge based on information there is necessarily an abstraction process associated. Such an abstraction has to face the problem of coherence or consistence of the possibly different individual interpretation processes. At any case, it must be taken care that there are no incompatible views merged.
- For information based on knowledge emphasizes this very *performance* of an interpretation, i.e. the actual usage of knowledge (items) for a considered action. The corresponding question remains namely in which sense we confine to the abstraction necessarily incorporated in the knowledge item.

As we will see both relationships are actually considered in practice. While the first one especially emphasizes the *generation* process, the latter rather stressses the *usability* of previously compiled information under given circumstances.

4 Conceptualizations of Information

The philosophical foundation has provided the epistemological background which leaves open the question of subsequent specifications. It is the task of the sciences to establish specifications that are consistent with the philosophical considerations and concretize aspects for their special purpose. In the following section we will investigate the contexts of such realizations.

4.1 Conceptualizations of Information in Organizational Units

In the context of business organizations [70] provides an excellent summary of both Western and Japanese interpretations of knowledge. The book also contains an extensive bibliography of the literature. For another account of the usage of these concept see also Robert Dunham's column [31] in the first issue of KM Briefs and KM Metazine where he gives an introduction of an understanding of 'knowledge' as being used in business organizations. In this context knowledge shows essentially a *dynamic* behaviour and may in short be characterized by "knowledge as action" (see [70, p. 57f], [76]) according to the theory of cognition in [60]. This view of knowledge has already been characteristic in Ackoff's approach where knowledge is considered as an *application* of data and information.

We also find a wide-spread understanding of information in the spirit of [27] in the business commmunity: "information is data with meaning" (see, e.g., [85], [23]).

Nonaka and Takeuchi emphasize the difference between *explicit knowledge*, which can be articulated in formal language and transmitted among individuals, and *tacit knowledge* which is understood as personal knowledge embedded in individual experience and involving such intangible factors as personal belief, perspective, and values – a distinction originally made by Michael Polanyi in 1966 [70, p. viii and p. 59]. This distinction has became very influential in the sequel. Nonaka and Takeuchi stress that "the interaction between these two forms of knowledge is the key dynamics of knowledge creation in the business administration" [70, p. ix].

It is thus the challenge of an organizational enterprize to acquire, to record, and to transform tacit knowledge into a suitable form such it subsequently can be turned into a valuable resource and utilized for the business' purposes and not only just remains an individual proficiency.

Nonaka and Tacheuchi summarize their study on the usage of information and knowledge in organizational structures and consider

> ...knowledge as a dynamic human process of justifying personal belief toward the 'truth'. [70, p. 58]

This is the viewpoint of what in enterprises is oftenly understood as "knowledge management". It acknowledges knowledge as a valuable but oftenly unarticulated and thus unexploited source. This view especially emphasizes the dynamic process of knowledge *compilation* being a central challenge of organizational units.

However, although these citation share a common view, these concept mostly appear to be ad-hoc definitions though and lack a profound foundation. This has already been mentioned in the survey in [86]

> Not only are the *definitions* of the three entities vague and imprecise; the relationships between them are not sufficiently covered.

The relation between data, knowledge, and information is often considered as constituting a hierarchy with data at the bottom, followed by information, and finally knowledge at the top (see [86, p. 3]). Stenmark then continues to critizise

> This image holds two tacid assumptions; Firstly, it implies that the relationship is asymmetrical, suggesting that data may be transformed into information, which, in turn, may be transformed into knowledge. However, it does not seem to be possible to go the other way.

One interesting point raised by [68] is the close relationship between *knowledge* and *abstraction* the latter being understood in contrast to *detail*.

> One of the most important characteristics of knowledge is **abstraction**, the suppression of detail until it is needed [...]. Knowledge is **minimization** of information gathering and reading – not increased access to information. Effective knowledge helps you eliminate or avoid what you **don't** want. Such abstraction also enables you to make judgments in a variety of situations, to generalize.

Consideration of that kind led to the notion of *information economics* [59] which is especially concerned with the study of the tangible value of information holdings to business enterprises exhibited by *information mining*.

It is worth mentioning that 'knowledge' in this field is considered essentially as a *process* that constitutes a challenge and demands a task. It thus shows a dynamic behaviour. The challenge is to exploit all (tacid) knowledge acquired by members of an organizational unit to the benefit of the organizational unit itself, and the task is the compilation of (individual) knowledge into a suitable generic form such that it can be utilized as a valuable resource. This compilation into a generic form may require some adjustment which is considered as abstraction.[4]

All these view share one common point that may be charactzerized as follows. Members of organizational units generally accumulate valuable personal experience or belief ("tacit knowledge") which are not entirely subjective and yet not fully objective [75]. As this constitutes a valuable source for the overall interest of the organizational unit, it is a challenge for the unit to "leverage the tacid knowledge of its members" ([86, p. 8]) and to transform this individuated form of organizational knowledge — which so far is only present in distributed form and only available to individuals — into a form of organizational knowledge that is available for the organizational unit itself ("explicit knowledge"). It is only the explicit form of knowledge that admits systematic exploitation on the organizational level and can be viewed as a secured valuable possession, i.e., that is uniformly accessible to each member of the unit and allows specific processing beyond individual awareness. This is the kernel of exploitation in the context of so-called "knowledge management". It may be summarized as follows: It is the challenge for every organizational unit

- to access individual experience ("tacid knowledge")
- to collect it in a form available for the unit ("knowledge management")
- to extract the valuable kernel out of these ("abstraction")
- to verify its contents ("true facts")
- to represent these in a form ready for exploitation ("explicit knowledge").

These conceptualization of 'data', 'knowledge', and 'information' so far reflect the overall aim of organizations interest. Emphasize is mostly on the *generation* and *aggregation* of knowledge by focussing on the *process* of abstraction whose *result* is in the light of the provious section necessarily associated with 'knowledge'. This is in some contrast to epistemic interests where emphasize is not

[4] This abstraction may only consist of a weak form as in Wiig [92] who defined information as "...facts organised to describe a situation or condition".

on problem of the generation and aggregation of knowledge but on knowledge as the disposable conceptual basis for personal orientation in and inter-personal communication on aspects of the world.[5]

4.2 Conceptualizations of Information in Information Science

There is a different understanding of 'knowledge' and 'information' to be observed in the information science and information technology communities on the one hand and the business-related approaches on the other hand. It was mainly in the eighties of the last century when the information science community in a wide-spread and detailed debate tried to become aware of the meaning of the fundamental concepts being constitutive for its field and to provide the conceptual ground for their self-understanding. In was mainly at that time when a discussion on foundational issues resulted in a detailed and carefully performed analysis of the so-far less specified concepts. As in the previous section, we again review some characterizations that have contributed to an overall understanding of the concept in the community and that appear to be fundamental for our intention as well.

This review, however, will beyond a mere compilation of more or less unrelated aspects at the same time provide a systematization of the so far scattered aspects of the concept of information. In this sense the different aspects are all meant to contribute to a faceted view of a whole.

Information as difference. The most abstract specification of the concept of information may be found in [11] (cf. also [10, p. 428] for some earlier considerations). According to Bateson

Information is [...] a difference that makes a difference. (1)

Although this fundamental characterization seems to meet the very kernel of a *conceptualization* of 'information', Bateson subsequently investigates possibilities for a universal definition and looks for a physicalistic foundation for a theory of information. We do not follow the overall assumption of Bateson, we do, however, think that Bateson has indeed identified the most abstract and

[5] It should be mentioned, though, that there are also understandings in the field of organizational units that not really conform with this view. Choo in "The Knowing Organization" [22, p. 62] for examples suggests an understanding of information being a change in the individual's state of knowledge and a capacity to act. This is a view that rather fits into an understanding as developped in the context of epistemological studies.

It may be annotated that even a radically different view has been suggested by Tuomi [89],[90]. He proposes that data emerges as a result of adding value to information, which in turn is knowledge that has been structured and verbalised (see also [86] on this topic). In the consequence there is no "raw" data. This is perfectly in concordance with the vision of an organizational unit to *store* the tacid knowledge for better exploitation.

general *description* of what 'information' is. As a consequence we separate his uniform description from the physicalistic presumption he is committed to.

The task remains to re-interpret this characterization in a suitable setting. Especially, a necessary concretization has to specify in what shape differences occur and in which granularity 'differences' may be recognized as such be it a person or a system, and especially what their content is, i.e. what is transmitted and what is the effect caused by it. In other words the *nature* of a 'difference' has to be clarified in order to make it an object of a theory of information.

Information as a process. Losee [55] tried to unify different approaches of conceptualizations of information and proposed a "domain-independent" view of information. This explicitly includes such different fields as physics and various models of human behavior. The unifying principle upon which his theory is based is found in an assumption about the *origin* of information. Losee claims that the fundamental phenomenon of information which underlies most other conceptualizations of information essentially depends on an underlying process (see [55, p. 258]). This is expressed in Losee's fundamental principle: "All processes produce information."

Information in turn must then be viewed as the result of a generation process: "the values within the outcome of any process". At first hand this makes information a dependent concept and hence *process* the primary concept to be studied for a general characterization of information. Accordingly, the focus is shifted from information itself to the conditions of its generation: the ability to "produce an effect" ([55, p. 259]).

One might be attempted to allude this notion of a process to the essential relationship to possible actions according to Peirce. But this line of thought is not pursued, and Losee looks for some other kind of justification. From a systematic point of view this approach poses several problems. It relates information to an understanding of 'process' with emphasize given on the detectable results 'value' and 'outcome'. Moreover, as long as these concepts are not proven to be more basic than the definiendum 'information', the definiton remains incomplete and rather results in an unnecessary inflation of basic concepts which are not obviously 'prior' from a systematic point of view.

Losee has been aware of this. In conformity with the etymological roots of 'information' Losee especially hints at the substring 'in—form—'. This should provide a justification for the intended specification of 'outcome' as 'characteristics in the output' (see [55, p. 256]) which results in the following statement:

Information is [...] the values of characteristics in the processes' output. (2)

This formulation, however, just extends the list of unspecified terms that have to be clarified by 'characteristics' and 'output'. Losee promotes his approach by presenting the prototypical models for such processes namely mathematical functions and causal mechanisms in physics. Two remarks should be pointed out.

Firstly, these notions already hint at some rather technical treatment of 'process' — and consequently of the concept of 'information' as well due to the dependencies established in definition (2). In this context 'value' undergoes a technical interpretation as a "variable returned by a function or produced by a process" which at the same time differentiates between processes and functions the latter being a subspecies of processes in a technical environment (see [55, p. 267]).

Secondly, definiton 2 seems to neglect that at "both ends of the channel cognitive processes occur" ([17, p. 195]). Especially this second limitation has been felt by Losee as well. To overcome this restriction — which would essentially undergo his attempt to establish a discipline independent definition of information — he tries to extend his procedural characterization of information to cognitive and communication processes as well. According to his approach, this amounts to interpret the 'characteristics' in a given field. The conceptualization of 'process' outside the technical areas including mathematics, however, remains altogether vague and is mainly inspired by his prototype models enriched by epistemic considerations. It is sketched like 'knowledge' — attributed as *justified* and *true* 'belief' — may be interpreted as *results* of processes by corresponding processing abilities of the mind ([55, p. 265ff]). This may be viewed as an attempt to demonstrate how domain-specific extensions like 'meaning' may be integrated into his approach to demonstrate its universality.

In general, his analysis lacks an adequate account on the very cognitive processes beyond a study of perception and its interpretation seen as a signal processing cognitive ability. Just in the contrary, capturing issues of a conceptual approach such as 'usefulness' is rather described as a *limitation* of a generic definition, since it restricts the "domain of discussions [...] to cognitive processes" ([55, p. 257]) and thus rather prevents a generic definition that should imply universal applicability. Losee even demands that domain-specific definitions should be set up in compatibility with his definition and thus constitute mere concretizations of his general description [55, p. 257].

There are aspects that make Losee's considerations important for our context though. In the light of Bateson's abstract specification ("a difference that makes a difference") Losee's definition of 'information' as inherently being the result of a process may be interpreted as a specification of that what actually "makes" a differences: it is caused by an underlying process — and we may add: driven by purpose and interest! In this sense Losee's principle may be re-interpreted as contributing a concretization for one facet of Bateson's abstract statement — and this is actually the point in Losee's approach we are interested in.

Information as transformation. The following statement from the influential view presented in [12] may be considered as to provide a further specification of (1). In our point of view it contains a partial answer to the question *on what* information makes a differnece and thus contributes to a more detailed understanding of the concept of information.

$$\text{Information is that which is capable of transforming structure.} \qquad (3)$$

According to a remark by Silvio Ceccato, the concept 'structure' in statement (3) is to be understood as a universal category in the philosophical understanding of the concept ([12, p. 198]). This might not have been the original intention of the authors, since they added a remarks expressing that the philosophical status of the concept 'structure' would not affect the argument. As we will see lateron this is absolutely right. This comment may rather be viewed as an expression of the authors' suspect that a systematization of this concept might be possible – and even desirable. We suspect that the contextualization of structure as a philosophical term has prevented the concept of information being studied in a formalized system where 'structure' indeed admits a concrete technical specification. Insofar the remark might be somewhat misleading in view of its consequences. It may already be annotated at this point that the work of Belkin and Robertsen has only been influential to the understanding of the concept of information in the context of information *science*. However, it has not led to a foundation of information *systems*.

Modification of knowledge structures. In the same direction of thinking a more intrinsic description is provided by [17, p. 197] who focusses on the nature of the transformation and the kind of structure that is altered:

Information is that which modifies [...] a knowledge structure. (4)

Altogether the abstract definition (3) of 'information' as "a difference that makes a difference" has been subject to a first concretization: Given a 'structure' (and we may add: which is represented in a suitable form) information is something external to this structure ("a difference") which may affect or change the internal disposition of this structure ("makes a difference") in a procedural sense. From this point of view it remains to consider three phenomena:

- The internal constitution of the structure.
- The nature or 'carriers' of the external influences that "make a difference".
- The kind of alteration of the structure after an affection by the external influences.

Information and knowledge. We have detected that information and knowledge *structures* are closely tied to each other. Moreover, *information* is defined as an alteration of a knowledge structure which makes *knowledge* prior to *information* from a systematic point of view. The definition of *knowledge* and *information* in [18, p. 131] acknowledges this very fact and focuses exactly on the dependency between these concepts.

knowledge is a linked structure of concepts. (5)

information is a small part of such a structure. (6)

Information and data. Another definition is based on a classic article by Richard Mason that relates information and data which — according to the philosophical commitment — are given as *signs* ([58]; see also section 3.2):

Information can be viewed as a collection of symbols [...] (7)

with the necessary addition "which has the potential to alter the cognitive state of a decision maker." This emphasizes the aspect what *carries* information complementing Losee's process-oriented approach and at the same time hints on specific settings that must be given to turn data into information: pure data viewed as (syntactically) organized signs are only considered as information in a suitable context, i.e. endowed with an interpretation providing meaning.

Information and meaning. The boundedness of information to interpretations (thus constituting meaning) is especially expressed in

> Something only *becomes* information when it is assigned a significance, interpreted as a sign, by some cognitive agent. [29, p. vii]

It should be noted, though, that this very view of the concept of information is critized by [29] as connecting information with meaning in a misleading way.

This definition relates 'information' to a state where "possible actions" are most desirable ("is assigned a significance"); in the light of the philosophical background we are committed to, the explicit relation to a decision maker just (rightly) links the concept of 'information' to a "pragmatic situation". Moreover, the definition expresses again that *signs* (symbols) are the carriers of information. However, the notion of an agent seems to be too restricted for a definition of 'information', since there are certainly other situation imaginable where a need for a subsequent action requires support. Altogether, the definition certainly points at some crucial situation in which the notion of 'information' plays a central role but restricts its scope unnecessarily.

5 Formalizations of the Concept of Information

We have shown that logic in the sense of Frege and Tarski has been designed as a semantic theory and rightly abstains from a pragmatic dimension. This is in full accordance with the philosophical tradition where 'truth' and 'proof' are thought of as being

- independent of the persons that demonstrate it,
- insensitive to their context,
- everlasting.

In contrast to this information essentially relies on the pragmatic aspect. This implies that a model for the situation in which information occurs along with a user model has necessarily to be included in a justified formal theory of information. Such a theory is out of sight though.

In this section we study formalizations of (aspects of) conceptualizations of 'information'. We then summarize what clarifications have already been made for a future comprehensive theory of information that is different from what is understood by 'information theory'.

5.1 Shannon's Theory of Information

The first section is be devoted to Shannon's theory of information ([82], [83]) which might justly called a break-through for a theory of dealing with a concept of 'information'.[6] Shannons's theory of information, however, — although a very early contribution to conceptualizations of 'information' in the light of the history of the development of the concept — aims at another direction than we are interested in. It is not a theory of communication in the sense of interpreting signs or selecting a subset of a given material for inspection according to an information request. Instead, it might rather be considered as a theory of transmission — as opposed to a theory of communication in the semiotic context — which reduces the concept of information to the decrease of uncertainty.

> The fundamental problem of communication is that of reproducing at one point either exactly or approximately a message selected at another point. Frequently the messages have meaning; that is, they refer to or are correlated according to some system with certain physical or conceptual entities. These semantic aspects of communication are irrelevant to the engineering problem. The significant aspect is that the actual message is one selected from a set of possible messages. [82, p. 379]

Shannon's theory of information thus explicitly excludes meaning in a Frege-Tarski-style and may rather we viewed as a theory of reduction of uncertainty in a situation in which a *given and known* set of alternatives together with corresponding probabilities is given. This means that we have a sort of a 'closed world' setting. The problems remains to get support for the selection of one of the alternatives. In this sense Shannon's theory may also be characterized as a theory of uncertainty under given (and known) circumstances.

The approach of Shannon (and Weaver) specializes and focusses of just some aspect of a concept of information — and it is its merrit to make this view subject for a fully developped theory in the sense of the (natural) sciences.

[6] It should be annotated though that — as usually in the development of science — there have been predecessors. Nyquist [71] already studied the limits of a telephone cable for the transmission of 'information'; Hartley [40] developed a conceptualization of information based on "physical as contrasted with psychological considerations" for studying electronic communication. Shannon also did not provide a *definition* of information but presented a *model* for a *theory* of information.

What Shannon essentially brought into the model is noise. See also [55] for a short survey of the development of 'information theory'.

Once understood, there have been more definitions (or rather explanations of the ground terms of this theory) in the same spirit in the sequel.[7] They all seek to capture the idea of the engineering aspect of information in the sense of Shannon more precisely. Such a definition that seems to meet the spirit of Shannon's intention in a more specific sense is provided by [61]: "Information is the content of the energy variations of the signal."

It has been noted, however, that aspects of this approach to reduce uncertainty may also be interpreted in the context of classical information retrieval: the fact that presenting more material to the person asking for information may well aggravate uncertainty [80]. But this remark may apply to information systems that do not rank documents according to their (estimated) retrieval value. More elaborate information systems that calculate a retrieval value status for the appropriateness of the documents for the query (or the information need, respectively) and only present the n highest ranked documents to the person requiring the information may not really be affected by this remark.

We agree with [57] that Shannon developped a theory for some subpart of a more general understanding of information insofar all sorts of information have the ability to reduce uncertainty. So Shannon's approach is far too restricted to be used for a basis of a conceptualization of information that may serve as a foundation for concepts of information as used in the design and modeling of information systems. This has already been pointed out by [17, p. 195] who noted that at ''both ends of the channel ... cognitive processes occur".

A review of understandings of the concept of 'information' that makes a claim on a certain completeness must not neglect these processes. To cope solely with the aspect of reducing uncertainty would mean to cut of a vivid tradition of conceptualizations of 'information' as we have shown. In some sense it is just on contrary the *increase* of certainty what information systems aim for. So it is our task to exhibit these cognitive processes.

5.2 Semantic Information

There are other conceptualizations of 'information' found in (the exact) science(s). The notion of semantic information has been brought up by the technical report of Bar-Hillel and Carnap [8] continuing some thoughts of Popper [77] and [78]; see also [9] and [48]. It is inspired by Shannon's approach to information and attempts to weave probability into a logical setting. Hintikka's pioneering paper of semantic information [42] presents a theory of information in the shape of a logical interpretation of probabiliy as opposed to a statistical (or frequency-based) theory of probability that examines what happens in the long run in certain types of uncertainty situations. A logical interpretation instead focuses its interest on possibilities to distinguish alternatives by means of their formal expressions of the logical language. The difference is characterized as follows:

[7] For a survey and an account on the development of ideas as well as more recent results in this area cf. [4].

> It is completely obvious that the sentence $(h\&g)$ can be said to be more informative than $(h \vee g)$ even if we know that both of them are in fact true.
>
> Semantic information theory accordingly arises when an attempt is made to interpret the probability measure p that underlies [inf(h)=–log p(h)] as being this kind of 'purely logical' probability. [42, p. 5]

An example of this 'purely logical' probability is provided by the following: Let K be the number of possible alternatives and consider a statement

$$h = C_1 \vee C_2 \vee \ldots \vee C_n$$

where the *width* $w(h)$ of h is $w(h) = n$. Then the 'logical' probability is

$$p(h) = \frac{w(h)}{2^K}$$

which results in the *information* of h (cf. Shannon's theory of information) as

$$\inf(h) = -logp(h) = -log(\frac{w(h)}{2^K}) = K - logw(h).$$

Fred Dretske's work [29] motivated and initiated a series of further developments in 'semantic information theory'. Continuing in this line Barwise and Perry [6] essentially expanded the scope and presented a theory which proved to be very influential in the sequel. Their work has been the starting point of Devlin's study of the concept of information [30], and subsequently Barwise and Seligman [7] also continued in this line of research.

A theory of information being developped as a logical theory necessarily inherits — according to the semantic nature of logic after Tarski [88] — its semantic status. This so far explains the denotation. Semantic information in this sense does not constitute an original understanding of information but rather suggest a treatment of the concept of information inside a logical framework.

The separation of information from necessary interpretation on the other hand allows to study its usage, its properties as pure carriers, and its functional behaviour. This is the program of [29] which is characterized as follows

> Once this distinction [of information and meaning] is clearly understood, one is free to think about information (though not meaning) as an objective commodity, something whose generation, transmission, and reception do not require or in any way presuppose interpretive processes. One is therefore given a framework for understanding how meaning can evolve, how genuine cognitive systems [...] can develop out of lower-order, purely physical, information-processing mechanisms.[...] The raw material is information. [29, p. vii]

The study of the functional phenomenon of information is the main task of the theory of information *flow*. For an account of recent research in this field see, e.g., [34], [35], [45].

5.3 Algorithmic Information Theory

Although certainly inspired by the basic considerations of Shannon's theory of information, Gregory Chaitin [21], Ray Solomonoff, and Andrej Kolmogorov (see, e.g., [53]) developed a view of information different from that of Shannon. Rather than considering the statistical ensemble of messages from an information source, algorithmic information theory focusses on individual sequences of symbols.

More precisely, the algorithmic information content or *Kolmogorov complexity* $H(X)$ of a string X is defined as the length of the shortest program p on a universal Turing machine (UTM) U producing string X. The field of algorithmic information theory is then devoted to the study of structures and phenomena of this concept.

As a subfield of mathematics, algorithmic information theory is completely unrelated to the concept of information used in information science or in the field of information retrieval. Algorithmic information theory is rather developped as an abstract theory of words and problems related to coding, complexity issues and pattern recognition. Hence we will not go into details of this theory in the sequel.

5.4 The Fundamental Equation of Information Science

There have been attempts to formalize conceptualizations of 'information' in information science.[8] It seems inevitable that certain reductions of the conceptual complexity of a socio-cultural understanding of the primary concepts has to go with such attempts. The semantic conception of 'truth' does definitely also not capture all facets of the understandings of the concept in real life! Instead, it is specifically tailored to capture the fundamental properties that are needed to build mathematics on it. All approaches towards a formalization of 'information', however, are far from constituting a formal system in the spirit of the theories described in this section — and a respective theory is out of sight at the moment.

The most promising starting point in view of admitting possible specifications of the basic symbols in terms of the cognitive sciences is certainly the famous 'fundamental equation' of information science which has been discussed in [16] and succesively refined until reaching its famous form in [18]:

$$K[S] + \Delta I = K[S + \Delta S] \tag{8}$$

This equation

> *...states in its very general way that the knowledge structure $K[S]$ is changed to the new modified structure $K[S + \Delta S]$ by the information ΔI, the ΔS indication the effect of the modification.* ([18, p. 131])

[8] See in particular Mizzaro's theory of information in [63], [64] which also provides an interesting approach. In this paper, however, we will rather concentrate on some other line.

Brookes has already attached some background ideas and annotations to statement (8) (cf. [18, p. 131]). In the course of his exposition he explains some principles behind this formally condensed statement. Especially, he points out that this equation must not be misunderstood as a simplification but as a *representation* of more complex insights. In this sense it is rather meant to transscript some general insights into the 'nature' of the concepts of knowledge and information.

Brookes' 'fundamental equation' provides the most abstract *formal* specification of the interaction of data, information, and knowledge and has experienced a broad acceptance in the information science and retrieval community. Our considerations are meant to provide a conceptual clarification of the significance of the fundamental equation beyond the original background along with a guideline for a realization of the rather abstract and merely denotational statement (8). The respective clarifications are at the same time the first step for any further modelling insofar they determine the abstract properties of and interrelations between the concepts that must be modelled. This will result in a specification of the *principles*[9] behind (8) in a theoretical framework.

5.5 Principles Ruling Information

In this section we summarize the properties along with the functionalities that have to be modeled by a theory of information. The principles (1) to (7) will constitute a system of prerequisites for any subsequent theory of information that claims to capture the essentials of the field.

(1) Information is [...] a difference that makes a difference.
(2) Information is [...] the values of characteristics in the processes' output.
(3) Information is that which is capable of transforming structure.
(4) Information is that which modifies [...] a knowledge structure.
(5) Knowledge is a linked structure of concepts.
(6) Information is a small part of such a structure.
(7) Information can be viewed as a collection of symbols.

A few remarks to these are meant to summarize and complement the longer expositions given in the previous section. In addition, their relation to the 'fundamental equation' will be explained in short:

(1) denotes the overall characterization along with the functional behavior of information. It refers to the Δ-operator in 8.
(2) expresses that there is a process involved whose *results* ΔI are the constituents of information. This is the part *information retrieval* especially focuses on.

[9] We called these basic specifications *principles* thus avoiding the wording *axioms* because we are well aware that in light of modern logic an axiomatization would require a formal language along with a syntax to formulate the axioms in.

(3) specifies the abstract 'difference' as a transformation process resulting in $K[S + \Delta S]$. At the same time it hints at some regularities ('structure') that must be present to impose influence upon.
(4) shows on what information causes effects, namely on knowledge (structures): $K[S]$
(5) characterizes the necessary internal constitution of such background *structures* (i.e. nothing being amorph) and especially provides a further condition on 'knowledge' (structures) to be able to admit effects at all.
(6) relates the internal constitution of information units to the internal constitution of knowledge structures. It states that certain compatibility conditions must be fulfilled in order to be able to impose effects.
(7) specifies the internal constitution of the *carriers* of information. Moreover, it gives the contents which in principle would admit a formal treatment. According to the semiotic approach these must be *signs.*

We have shown that statement (8) indeed incorporates — in coded symbolic form — fundamental determinants of the concept of 'information' along with its relations to the concepts necessarily connected with it as studied in the previous sections of this paper. In this sense this paper contributes to the general research program initiated by Brookes [18, p. 117] when stating that

> the interpretation of the fundamental equation is the basic research task of information science.

A systematic analysis in the spirit of Peirce's theory of concepts — namely to exhibit possible actions involved therein — thus led to an abstract specification of fundamental properties implicitly involved therein. The principles we found complement the formal statement and the abstract discussion as well. They provide a pre-formal intermediate layer between pure philosophical considerations on one hand and formalizations on the other hand which results in a new dimension in this field.

6 Conclusion and Outlook

Reviewing the approaches to determine the concept of information, we found a lot of scattered characterizations. Whereas some seem less convincing or guided by some other interest, most of these do meet some single facets of the point but are at same time insufficient and lack a clear and solid comprehensive foundation of the concept of information. In this situation the present paper indicates an approach for a promising systematization on the basis of the semiotical approach in the succession of Peirce and guided by the 'fundamental equation' of Brookes. In accomplishing the basic steps we have provided a well-founded and solid perspective for future work in this direction. It is to be hoped that a foundation will finally evolve that may cope with foundations of the concept of 'truth' and 'proof' as provided for logic and thus for the whole mathematics and especially may constitute a framework for the modelling purposes in computer science.

References

1. Computing Curricula 2001 Computer Science. Final Report. Available at http://www.acm.org/sigcse/cc2001, 2001.
2. Ackoff, R.L.: Transformational consulting. Management Consulting Times, Vol. 28, No. 6 (1997)
3. Alavi, Maryam, Leidner, Dorothy E.: Knowledge Management and Knowledge Management Systems: Conceptual Foundation and An Agenda for Research. MIS Quarterly, March 2001, pp. 107–136.
4. Arndt, Christoph: Information Measures. Information and its Description in Science and Engineering. Springer-Verlag, Heidelberg, 2001.
5. Ayer, A.J.: The Problem of Knowledge. Macmillan, London, 1956.
6. Barwise, K. Jon, Perry, J.: Situations and Attitudes. MIT Press, Cambridge, MA, 1983.
7. Barwise, J.; Seligman, J.: Information Flow: The Logic of Distributed Systems. Cambridge University Press, Cambridge, 1997.
8. Bar-Hillel, Yoshua, Carnap, Rudolf: An Outline of a Theory of Semantic Information. Research Laboratory for Electronics, MIT, Cambridge, 1952, Technical report no. 247.
9. Bar-Hillel, Yoshua, Carnap, Rudolf: Semantic Information. British Journal for the Philosophy of Science, Vol. 4 (1953), pp. 147–157.
10. Bateson, G.: Steps to an ecology of mind. Paladin, St. Albans, Australia, 1973.
11. Bateson, G.: Mind and nature: a necessary unit. Bantam Books, 1980.
12. Belkin, Nicholas J., Robertson, Stephen E.: Information Science and the Phenomenon of Information. Journal of the American Society for Information Science, 1976, pp. 197–204.
13. Belkin, Nicholas J.: The Cognitive Viewpoint in Information Science. Journal of Information Science, Vol. 16 (1990), pp. 11–15.
14. Boole, George: An Investigation of the Laws of Thought, on which are Founded the Mathematical Theories of Logic and Probabilities. Walton & Maberly, London, 1854.
15. Boole, George: The calculus of logic. Cambridge Dublin Math. J. Vol. 3 (1848), pp. 183–198.
16. Brookes, Bertram C.: The Fundamental Equation in Information Science. Problems of Information Science, FID 530, VINITI, Moscow, 1975, pp. 115–130.
17. Brookes, Bertram C.: The Developing cognitive viewpoint in information science. In: de Mey, M. (ed.): International Workshop on the Cognitive Viewpoint. University of Ghent, Ghent, 1977, pp. 195–203.
18. Brookes, Bertram C.: The Foundations of Information Science. Part I. Philosophical Aspects. Journal of Information Science, Vol. 2 (1980), pp. 125–133.
19. Buckler, J. (ed.): Philosophical writings of Peirce. Dover, New York, 1955.
20. Carnap, Rudolf: Foundations of Logic and Mathematics. University of Chicago Press, Chicago, 1939.
21. Chaitin, Gregory J.: Algorithmic Information Theory. Cambridge University Press, Cambridge, 1990 (3rd ed.).
22. Choo, Chun Wei: The Knowing Organization. Oxforf University Press, New York, 1998.
23. Choo, Chun Wei, Detlor, Don, Turnbull, Don: Web Work: Information Seeking and Knowledge Work on the World Wide Web. Kluwer, Dordrecht, 2000.

24. Church, Alonzo: A Bibliography of Symbolic Logic. The Journal of Symbolic Logic, Vol. 1, No. 4 (1936), pp. 121–216.
Additions and Corrections to: A Bibliography of Symbolic Logic. The Journal of Symbolic Logic, Vol. 3, No. 4 (1938), pp. 178–192.
25. Dedekind, R.: Was sind und was sollen die Zahlen? Vieweg, Wiesbaden, 1888.
26. Deledalle, Gérard: Charles S. Peirce: An Intellectual Biography. John Benjamins, Amsterdam, 1990.
27. Denning, Peter J.: The IT Scholl Movement. Communications of the ACM, Vol. 44, No. 8, (2001), pp. 19–22.
28. Descartes, Rene: Discourse on the Method of Rightly Conducting the Reason. In: The Philosophical Works of Descartes. 2 vols. Translated by E.S. Haldane and G.R.T. Ross. Dover, 1931, Vol. 1, pp. 79–130.
29. Dretske, Fred: Knowledge and the Flow of Information. MIT Press, 1981.
30. Devlin, Keith: Logic and information. Cambridge Univ. Press, 1991.
31. Dunham, Robert: Knowledge in action: the new business battleground. KM Metazine, Issue 1, (1996), http://www.ktic.com/topic6/KMBATTLE.HTM.
32. Eliasmith, Chris: Dictionary of Philosophy of Mind. Available at http://www.artsci.wustl.edu/ philos/MindDict/index.html.
33. Fisch, Max H.: Introduction to Writings of Charles S. Peirce. Indiana University Press, Bloomington, 1982.
34. Floridi, Luciano: Outline of a Theory of Strongly Semantic Information. Submitted; also available at http://www.wolfson.ox.ac.uk/~floridi/pdf/otssi.pdf.
35. Floridi, Luciano: Is Information Meaningful Data? Submitted; also available at http://www.wolfson.ox.ac.uk/~floridi/pdf/iimd.pdf.
36. Frege, F.L. Gottlob: Begriffsschrift, eine der arithmetischen nachgebildete Formelsprache des reinen Denkens. Nebert, Halle, 1879
37. Gallie, W.B.: Peirce and Pragmatism. Penguin, Harmondsworth, 1952.
38. Gehlen, Arnold: Der Mensch: Seine Natur und seine Stellung in der Welt. In: Gehlen, Arnold: Gesamtausgabe, Rehberg, Karl-Siegbert (ed.) Vittorio Klostermann, Frankfurt am Main, 1993.
39. Goedel, Kurt: Über formal unentscheidbare Sätze der 'Principia Mathematica' und verwandter Systeme I. Monatshefte Math-Phys. Vol. 38 (1931), pp. 173–198.
40. Hartley, R.V.L.: Transmission of Information. Bell System Technical Journal, Vol. 7 (1928), pp. 535–563.
41. Hintikka, K. Jaakko J.: Knowledge and Belief. Cornell University Press, Ithaca, NY, 1962.
42. Hintikka, K. Jaako J.: On semantic information. In: Hintikka, K. Jaako J., Suppes, Patrick (eds.): Information and Inference. Reidel, Dordrecht, 1970, pp. 3–27.
43. Hookway, Christopher: Peirce. Routledge & Kegan Paul, London, 1985.
44. Ingwersen, Peter: Information Retrieval Interaction. Taylor Graham, London, 1992.
45. Israel, David, Perry, John: What is Information? In: Hanson, Philip (ed.): Information, Language, and Cognition. University of British Columbia Press, Vancouver, 1990, pp. 1–19.
46. Johnson-Laird, P.H., Wasow, P.: Thinking: Readings in Cognitive Science: an Introduction to the Scientific Study of Thinking. Cambridge University Press, Cambridge, 1977.
47. Kant, Immanuel: Kritik der reinen Vernunft. Riga, 1789.
48. Kemeny, J.G.: A Logical Measure Function. Journal of Symbolic Logic, Vol. 18 (1953), pp. 289–308.
49. Kemerling, Garth: Philosophy Pages. Available at http://www.philosophypages.com/index.htm.

50. Lenski, Wolfgang, Wette-Roch, Elisabeth: Metadata for Advanced Structures of Learning Objects in Mathematics. An Approach for TRIAL-SOLUTION. Available at http://www-logic.uni-kl.de/trial/metadata_v2-0.pdf
51. Lenski, Wolfgang, Wette-Roch, Elisabeth: Foundational Aspects of Knowledge-Based Information Systems in Scientific Domains. In: Klar, R., Opitz, O. (eds.): Classification and knowledge organization. Springer-Verlag, Heidelberg, 1997, pp. 300–310.
52. Lenski, Wolfgang, Wette-Roch, Elisabeth: Pragmatical Issues in Scientific Information Systems. In: B. Sanchez, N. Nada, A. Rashid, T. Arndt, M. Sanchez (eds.): Proceedings of the World Multiconference on Systemics, Cybernetics and Informatics SCI2000. IIIS, Orlando, 2000, pp. 242-247 (also in CD-ROM of SCI2000).
53. Li, Ming, Vitanyi, Paul: An Introduction to Kolmogorov Complexity and Its Applications. Springer-Verlag, Heidelberg, 1997 (2nd ed.).
54. Lindsay, P.H., Norman, D.A.: Human Information Processing. Academic Press, New York, 1977.
55. Losee, Richard M.: A Discipline Independent Definition of Information. Journal of the American Society for Information Science, Vol. 48, No. 3 (1997), pp. 254–269.
56. Luft, Alfred Lothar: Zur begrifflichen Unterscheidung von "Wissen", "Information" und "Daten". In: Wille, Rudolf, Zickwolff, Monika (eds.): Begriffliche Wissensverarbeitung. Grundlagen und Aufgaben. BI Mannheim, 1994, pp. 61–79.
57. Machlup, F.: Semantic Quirks in Studies of Information. In: Machlup, F., Mansfield, U. (eds.): The Study of Information John Wiley & Sons, New York, 1983, pp. 641–671.
58. Mason, Richard: Measuring Information Output: A communication Systems Approach. Information and Management, Vol. 1 (1978), pp. 219–234.
59. Mattessich, Richard: On the nature of information and knowledge and the interpretation in the economic sciences. Library Trends, Vol. 41, No. 4 (1993), pp. 567–593.
60. Maturana, H.R., Varela, F.J.: The Tree of Knowledge. Shambala, Boston, 1992.
61. Menant, Christophe: Essay on a systemic theory of meaning. Available at http://www.theory-meaning.fr.st/.
62. Mey de, M.: The Cognitive viewpoint: its Development and its Scope. In: de Mey, M. (ed.): International Workshop on the Cognitive Viewpoint. University of Ghent, Ghent, 1977, pp. xvi–xxxii.
63. Mizzaro, Stefano: On the Foundations of Information Retrieval. In: Atti del Congresso Nazionale AICAÕ96 (Proceedings of AICA'96), Roma, IT, 1996, pp. 363–386..
64. Mizzaro, Stefano: Towards a theory of epistemic information. Information Modelling and Knowledge Bases. IOS Press, Amsterdam, Vol. XII (2001), pp. 1–20.
65. Morris, Charles W.: Foundation of the theory of signs. In: International Encyclopedia of Unified Science, Vol. 1, No. 2, University of Chicago Press, Chicago, 1938.
66. Morris, Charles W.: Signs, Language, and Behaviour. New York, 1946.
67. Murphey, Murray G.: The Development of Peirce's Philosophy. Harvard University Press, Cambridge, 1961.
68. Murray, Philip C.: Information, knowledge, and document management technology. KM Metazine, Issue 2 (1996), http://www.ktic.com/topic6/12_INFKM.HTM.
69. Nagel, E.: The structure of Science. Hackett, Indianapolis, 1979.
70. Nonaka, Ikujiro, Takeuchi, Hirotaka: The Knowledge-Creating Company. Oxford University Press, Oxford, 1995.

71. Nyquist, H.: Certain factors affecting telegraph speed. Bell System Technical Journal, Vol. 3 (1924), pp. 324–346.
72. Parker, Kelly A.: The Continuity of Peirce's Thought. Vanderbilt University Press, Nashville, TN, 1998.
73. Peckhaus, Volker: 19th century logic between philosophy and mathematics. Bulletin of Symbolic Logic, Vol. 5, No. 4 (1999), pp. 433–450.
74. Peirce, Charles S.: Collected Papers of Charles Sanders Peirce, Vols. 1–8. Hartshorne, Charles, Weiss, Paul (vols. 1–6), Burks, Arthur W. (vols. 7–8) (eds.) Harvard University Press, Cambridge, Mass., 1931–1958.
75. Polanyi, Michael: Personal knowledge. Routledge, London, 1958, (corrected ed. 1962).
76. Polanyi, Michael: Tacit Dimension. Doubleday, Garden City, NY, 1966.
77. Popper, Karl Raimund: Logik der Forschung. Springer, Wien, 1934.
78. Popper, Karl Raimund: Degree of Confirmation. British Journal for the Philosophy of Science, Vol. 5 (1954), pp. 143–149. Correction ibid. p. 334.
79. Rapaport, William J.: How to Pass a Turing Test: Syntactic Semantics, Natural-Language Understanding, and First-Person Cognition. Journal of Logic, Language, and Information, Vol. 9, No. 4 (2000), pp. 467–490.
80. Roberts, N.: Social considerations towards a definition of information science. Journal of Documentation, Vol. 32, No. 4 (1976), pp. 249–257.
81. Sanders, Donald H.: Computers in Society. McGraw-Hill, New York, 1973.
82. Shannon, Claude E.: A Mathematical Theory of Communication. The Bell System Technical Journal, Vol. 27 (1948), pp. 379–423, 623–656.
83. Shannon, Claude E., Weaver, W.: The mathematical theory of communication. University of Illinois Press, Urbana, 1949.
84. Schroeder, F.W.K. Ernst: Vorlesungen über die Algebra der Logik (exakte Logik) Vol. 1, 2. Teubner, Leipzig, 1890, 1891.
85. Spek van der, Rob, Spijkervet, André: Knowledge Management: Dealing Intelligently with Knowledge. CIBIT, Utrecht, 1997.
86. Stenmark, Dick: The Relationship between Information and Knowledge. In: Proceedings of the 24th Information Systems Research Seminar in Scandinavia (IRIS 24), Ulvik, Hardanger, Norway, August 11–14. Department of Information Science, University of Bergen, Bergen, 2001. (CD-Rom)
87. Suppes, Patrick (ed.): The Structure of Scientific Theories. Univ. of Illinois Press, Urbana, 1977.
88. Tarski, Alfred: Der Wahrheitsbegriff in den Sprachen der deduktiven Disziplinen. Anzeiger der Österreichischen Akademie der Wissenschaften, Mathematisch-Naturwissenschaftliche Klasse, Vol. 69 (1932), pp. 23–25.
89. Tuomi, Ilkka: Data is more than Knowledge: Implications of the Reversed Knowledge Hierarchy for Knowledge Management and Organizational Memory. Journal of Management Information Systems, Vol. 16, No. 3 (1999), pp. 107–121.
90. Tuomi, Ilkka: Corporate Knowledge: Theory and Practice of Intelligent Organizations. Metaxis, Helsinki, 1999.
91. Turing, Alan: On computable numbers, with an application to the 'Entscheidungsproblem'. Proc. of the London Math. Society, Series 2, Vol. 42 (1936), pp. 544-546.
92. Wiig, Karl M.: Knowledge Management Foundations: Thinking About Thinking — How People and Organizations Represent, Create, and Use Knowledge. Schema Press, Arlington, TX, 1994.
93. Wilson, T.D.: The cognitive viewpoint: its development and its scope. Social Science Information Studies, Vol. 4 (1984), pp. 197–204.
94. Zadeh, Lotfi: Fuzzy sets. Information & Control, Vol. 8 (1965), pp. 338–353.

Retrieval by Structure from Chemical Data Bases

Thomas Kämpke

Forschungsinstitut für anwendungsorientierte Wissensverarbeitung FAW
Helmholtzstr. 16, 89081 Ulm, Germany
kaempke@faw.uni-ulm.de

Abstract. Screening is an essential function of chemical compound data bases or combinatorial libraries. Two version of screening are distinguished: searching the data base for compounds that either match or contain some user-specified compound of interest, the query compound. Both versions of screening – structure query and substructure query – are always understood in some approximative sense. Similarity notions for molecular graphs are given here in terms of graph Voronoi regions. These lead to certain shortest path lists and shortest path matrices to which retrieval algorithms refer.

Keywords: Combinatorial library, compound data base, screening.

1 Introduction

Retrieval by structure in chemical data bases also called combinatorial libraries consists of finding all stored molecules that match some query structure approximately. Retrieval is posed in an approximate sense since approximate matches are always – and often even more than exact matches – of interest to the working chemist. The size of combinatorial libraries ranges up to several hundred thousand compounds which calls for similarity concepts that are efficiently computable.

The similarity issue is related to the diversity problem which amounts to finding a subset of a combinatorial library that bears much of the diversity of the original library. Maximizing diversity is thus reduced to minimizing similarity.

Similarity will be derived here from graph Voronoi regions and their shortest path trees. In addition, searching for all superstructures of a query is supported by distance methods.

This work is organized as follows. Molecular graphs are introduced in section 2. Distance-based graph notions for similarity and corresponding retrieval algorithms will be stated in section 3. Section 4 deals with a numerical similarity measure. Section 5 covers substructure retrieval. This amounts to finding all compounds that contain the query in an approximate sense.

W. Lenski (Ed.): Logic versus Approximation, LNCS 3075, pp. 106–119, 2004.

2 Representation and Retrieval Problems

2.1 Molecular Graphs

Each molecule is denoted by a molecular graph or chemical graph $G = (V, E, \gamma)$. This is an undirected graph $G = (V, E)$ with vertex set V, edge set E and vertex label function $\gamma : V \to \{C, O, H, Na, Cl, \ldots\}$. The vertices indicate the atoms of the molecule, the edges indicate bonds and the vertex labels indicate the atomar types. The atomar types which appear in a molecular graph G are denoted by $Type(G)$.

Molecular graphs of organic and many other substances are planar, but in exceptions like epoxies and zeolits molecular graphs are not planar [8]. This means that no crossing-free drawing exists. Violations of planarity will be infrequent and whenever a molecular graph is planar it will be drawn without edge crossings.

Two chemical graphs are isomorphic if the unlabelled graphs are isomorphic, which means that they admit a neighbourhood preserving one-to-one mapping, and the neighbourhood preserving mapping preserves vertex labels. Isomorphism serves as a starting point for deriving similarity concepts.

Any two vertices $v, w \in V$ of a molecular graph are connected by a path whose length is the number of its edges. The length of a shortest path between two vertices is denoted by their distance $D(v, w)$. The number of edges which meet in a vertex is its degree. The degree of any vertex is at most eight in molecular graphs. Thus, molecular graphs including all non-planar ones fall into the class of linear graphs which are those graphs whose edge numbers are linearly bounded in the number of vertices; $|E| = O(n)$.

2.2 Retrieval

Retrieval requires comparison of one structure, the query structure, against all structures stored in a combinatorial library and reporting of all approximate matches. The typical size of a combinatorial library may be several hundred entries but this size can easily increase to several hundred thousands. Each structure may consist of an unbounded number of atoms with few hundreds of atoms being no exception.

A variant of the approximate retrieval problem is substructure retrieval. Therefore, all library compounds are to be retrieved which are a superstructure of the query structure, i.e. which contain the query. Again, this inclusion relation is understood in an approximate sense. The construction of the query compound is left to the user in all cases.

2.3 Related Work

A plethora of so-called indices has been suggested for the description of molecular similarity [5], [20]. An index is a real number assigned to each compound with a smaller distance corresponding to higher similarity. In this respect structure retrieval is related to case based reasoning which can be considered to consist of the four components *vocabulary*, *case base*, *similarity measure* and *solution adaptation* [18].

The assignment of numbers by molecular indices is typically degenerate meaning that different compounds receive the same value. One of the earliest indices was the Wiener index which amounts to the sum of the lengths of shortest paths between all vertex pairs [11]. Other indices are the Balaban index and the Hosoya index [19] which is based on the adjacency matrix of the molecular graph. Similar indices are based on the cyclic structure of a molecule [6].

Instead of encoding each compound by one value, the compound may be encoded by matrices such as the adjacency matrix, distance matrix, the detour matrix etc. These exhibit a high degree of non-degeneracy but are quadratic in the number of atoms of the compound.

In another line of approaches, similarity is addressed by descriptors or features of almost any kind [10]. A descriptor amounts to the occurrence of certain substructures, bond types etc. Substructure consideration is motivated by chemical activity often being attributed to certain parts, the so-called functional groups of a molecule. The number of descriptors is variable but should be identical for all compounds of one library. The particular set of descriptors may remain undisclosed for certain data bases due to proprietary reasons. Similarity of compounds is measured by the distance between descriptors. Distance is understood as Euclidean distance, Hamming distance or the Tanimoto distance in case of binary descriptors, or one of the numerous variations thereof [9].

The most wide spread index seems to be Tanimoto index. This index is computed for any two compounds A, Q with binary descriptor vectors x_A, x_Q as $D(A,Q) = 1 - \frac{c}{a+q+c}$, where a is the number of 1's exclusive in x_A, q is the number of 1's exclusive in x_Q and c is the number of 1's common to x_A and x_Q. As an example, consider the description vectors $x_Q = (1,1,0,0,0,0,1)$ and $x_A = (1,1,1,0,0,0,0)$. The Tanimoto distance is $D(A,Q) = 1 - \frac{2}{1+1+2} = 0.5$, where $a = 1$, $q = 1$ and $c = 2$.

Similarity retrieval based on descriptors is a feature of chemical data bases such as Beilstein's CrossFire 2000 [3] and others [1], [4]. Another descriptor based system with several hundred features and (again) similarity assessment by the Tanimoto index is Carol, see [17]. Unless encoded by a descriptor, no topological information is considered.

3 Graph Similarity

3.1 Graph Voronoi Regions

Graph Voronoi regions are easily extended from planar geometry to undirected and directed graphs, see [7] for the latter. Here, a version for labeled graphs is required. The graph Voronoi region of a particular vertex or site vertex in a molecular graph is given by all vertices of other types which have no larger distance to the site vertex than to any other site vertex of the same type. To avoid trivial complications, the site vertex itself also belongs to its Voronoi region. The graph Voronoi region of a vertex p is formally defined by $V(p) = \{p\} \cup \{q \in V | \gamma(q) \neq \gamma(p)$ and $D(p,q) \leq D(p',q) \, \forall p' \in V$ with $\gamma(p') = \gamma(p)\}$.

Graph Voronoi regions always form a cover of the molecular graph. A sample of graph Voronoi regions is given in figure 1. Considering graph Voronoi regions

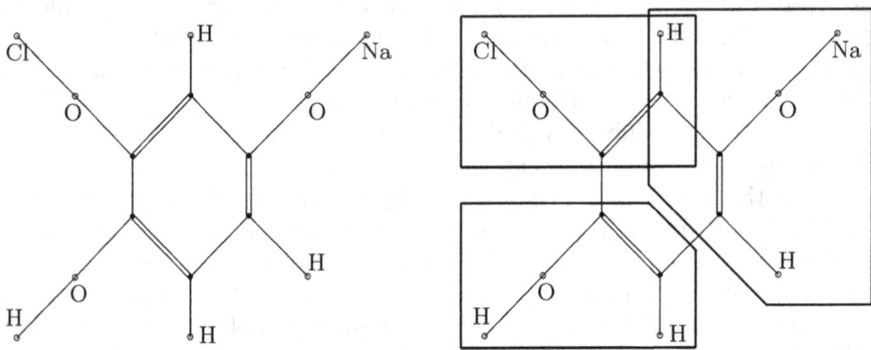

Fig. 1. Sample compound (left) and its overlay with the graph Voronoi regions of the three oxygon atoms (right). Multiple bonds are not considered for distance computations. The carbon and the hydrogen atom in the intersection of the two upper regions have the respective distances 2 and 3 from their site vertices.

provide for limiting the interaction between atoms as compared to considering all vertex pairs of two distinct types. This is achieved by lists of path lengths. The <u>shortest path list</u> or (γ_0, γ_1) shortest path list of a Voronoi region with site vertex p of type $\gamma(p) = \gamma_0$ is defined as increasingly sorted list of distances to all vertices of type γ_1 within the Voronoi region. Formally, the (γ_0, γ_1) shortest path list for Voronoi region $V(p)$ is given by $SPL(\gamma_0, \gamma_1) = SPL_p(\gamma_0, \gamma_1) = D(p, q_1), \ldots, D(p, q_{n(p)})$ where

1. $\gamma(p) = \gamma_0 \neq \gamma_1$,
2. $\{q \in V(p) | \gamma(q) = \gamma_1\} = \{q_1, \ldots, q_{n(p)}\}$,
3. $D(p, q_1) \leq \ldots \leq D(p, q_{n(p)})$.

Whenever the vertex type γ_1 does not occur in some Voronoi region $V(p)$, the shortest path list is set to be $SPL_p(\gamma_0, \gamma_1) = 0$.

Shortest path lists are illustrated for the compounds from figure 2 in the following table. The oxygen atom p refering to the last shortest path list is the lower left oxygen atom in both cases. Its Voronoi region contains two hydrogen atoms in the first case and only one in the second case.

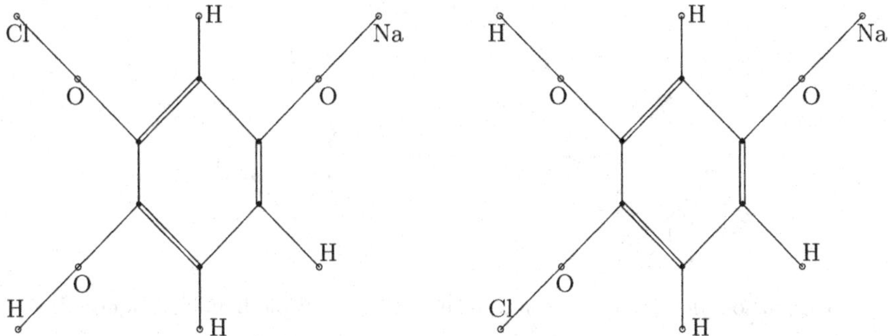

Fig. 2. The substance from figure 1 and a slight modification. Some graph Voronoi regions and corresponding shortest path lists are different.

	First compound	Second compound
$SPL(Na, Cl)$	6	7
$SPL(Cl, Na)$	6	7
$SPL(Na, O)$	1,5,6	1,5,6
$SPL(Cl, H)$	4,5,5,6	4,5,5,5
$SPL_p(O, H)$	1,3	3

Shortest path lists are now fused to lists of lists or second order lists. The list of shortest path lists for any pair of distinct vertex types (γ_0, γ_1) is given by $LSPL = LSPL(\gamma_0, \gamma_1) = LSPL^G(\gamma_0, \gamma_1) = (SPL_p(\gamma_0, \gamma_1))$ for all vertices p of type $\gamma(p) = \gamma_0$. The shortest path lists of any list of lists are sorted by increasing size and equally long lists are arranged in lexicographic order.

The list of shortest path lists for the (O,H) atoms and the (CL,O) atoms of the foregoing two compounds consist of the following entries.

	First compound	Second compound
$LSPL(O, H)$	3	3
	1,3	1,3
	3,3	3,3
$LSPL(Cl, O)$	1,4,5	1,4,6

3.2 Shortest Path Similarity

Molecular graphs are considered as similar if all their lists of shortest path lists are identical. Formally, two molecular graphs G_1, G_2 are shortest path similar or similar if $LSPL^{G_1}(\gamma_0, \gamma_1) = LSPL^{G_2}(\gamma_0, \gamma_1)$ for all pairs of distinct atomar types $\gamma_0, \gamma_1 \in Type(G_1) = Type(G_2)$.

Two isomorphic molecular graphs have identical Voronoi regions and, consequently, are similar but similar molecules need not be isomorphic as indicated by figure 3. The presence of only two types of atoms implies that there are only

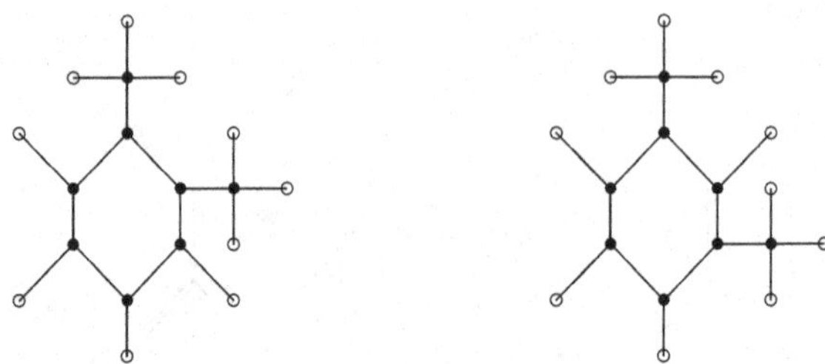

Fig. 3. o-xylene (left) and m-xylene (right). Multiple bonds are ignored.

two lists of shortest path lists. Each second order (hydrogen,carbon) list contains ten entries of "1". The second order (carbon,hydrogen) lists are as follows.

	o-xylene	m-xylene
$LSPL(C,H)$	0	0
	0	0
	1	1
	1	1
	1	1
	1	1
	1,1,1	1,1,1
	1,1,1	1,1,1

The two carbon atoms from the benzene ring which connect to the methyl groups have no hydrogen atom in their Voronoi regions. Thus, the shortest path lists of these atoms consist of zero entries.

3.3 Voronoi Regions and Shortest Path Lists

The graph Voronoi regions for all vertices and all their shortest path lists can be computed by a single run of a many to many version of the Dijkstra algorithm [15]. Each Voronoi region is therefore represented by its shortest path tree. This tree is rooted at the site vertex of the Voronoi region and specifies one shortest path to any of its vertices. The subsequent algorithm computes all Voronoi regions and all shortest path lists for all site vertices of a selected type γ_0.

VorList

1. Input γ_0, $G = (V, E, \gamma)$.
   Initialization. Set $S = \{v \in V | \gamma(v) = \gamma_0\}$,
   $\lambda(v) = 0$ and $site(v) = \{v\}$ $\forall v \in S$,
   $\lambda(v) = \infty$ and $site(v) = \emptyset$ $\forall v \in V - S$,
   $succ(v) = \emptyset$ $\forall v \in V$, $L = V$,
   $SPL_v(\gamma_0, \gamma_1) = 0$ $\forall \gamma_1 \neq \gamma_0$ and $\forall v \in S$.

2. While $L \neq \emptyset$ do
   a) Selection of $i = argmin_{l \in L} \lambda(l)$.
   b) $L = L - \{i\}$.
   c) List $SPL_v(\gamma_0, \gamma(i))$ receives entry $\lambda(i)$ at the tail $\forall v \in site(i)$.
   d) $\forall j \in L$ with $(i, j) \in E$ do
      if $\lambda(i) + 1 \leq \lambda(j)$ then
      i. $\lambda(j) = \min\{\lambda(j), \lambda(i) + 1\}$
      ii. $succ(i) = succ(i) \cup \{j\}$
      iii. $site(j) = site(j) \cup site(i)$.
3. Output graphs of $V(v)$, $\forall v \in S$, specified by shortest paths $(v = w_0, w_1, \ldots, w_m)$ in forward notation $w_{i+1} \in succ(w_i)$, $i = 0, \ldots, m-1$.
   Output shortest path lists $SPL_v(\gamma_0, \gamma_1)$ $\forall \gamma_1 \neq \gamma_0$ and $\forall v \in S$ with leading 0 removed in all lists with additional entry.

The value $\lambda(i)$ denotes the length of a best path found so far from any of the site vertices to vertex i. The successor list $succ(i)$ of any vertex i denotes the set of those vertices which lie immediately behind i on a best path found so far.

The forward notation of paths instead of the usual backward notation by the Dijkstra algorithm applies since all edges contribute equal values to the path lengths. The shortest path lists are increasingly sorted, because the minima of the λ function as computed in step 2(a) increase with iterations.

The foregoing algorithm can be used to compute shortest path lists without the graphs of the Voronoi regions. This is obtained by omitting the successor lists in the initialization and by omitting step 2(d) ii.

A straightforward implementation of the algorithm **VorList** runs in $O(n^2)$ time because all molecular graphs are linear. The bound can be improved to $O(n \log n)$, if the labels $\lambda(\cdot)$ are arranged in a heap. This heap is computable in time $O(n \log n)$, the minimum in step 2(a) can be retrieved in $O(\log n)$ and a constant number of updates in step 2(d) i can be made in $O(\log n)$ as well. Step 2 is iterated n times.

All second order lists can be computed according to the following procedure which simply runs algorithm **VorList** for all atomar types.

AllVorList

1. Input $G = (V, E, \gamma)$.
2. For all $\gamma_0 \in Type(G)$ do
   Compute shortest path lists $SPL_p(\gamma_0, \gamma_1)$ $\forall \gamma_1 \in Type(G) - \{\gamma_0\}$ and $\forall p \in V$ with $\gamma(p) = \gamma_0$.
3. Arrange shortest path lists to second order lists $LSPL^G(\gamma_0, \gamma_1)$ for all pairs of distinct values $\gamma_0, \gamma_1 \in Type(G)$.

Procedure **AllVorList** can be guaranteed to run in $O(|Type(G)| \cdot n^2)$ time and the bound reduces to $O(|Type(G)| \cdot n \log n)$ when the complexity of each execution of the second step is reduced to $O(n \log n)$.

The space requirement of all second order lists has a worst case bound of $O(n^2)$. This bound is quadratic only because of potential overlaps of the Voronoi regions. The size of all second order lists is $O(n)$ if all Voronoi regions are pairwise disjoint.

3.4 Screening

A library $\Lambda = (G_i)_{i=1}^N$ of molecular graphs $G_i = (V_i, E_i, \gamma_i)$ can now be readily screened for a query substance G. The molecular graphs G_i have $|V_i| = n_i$ elements. The set of all atomar types that actually occur in the library is indicated by $Type(\Lambda) = \cup_{i=1,\ldots,N} Type(G_i)$.

Any library can be preprocessed so that all second order lists of each molecular graph are available at run time. Then, only the second order lists of the query substance need to be computed. The remainder is mere list comparison. The resulting algorithm which screens the library compound by compound is as follows.

Screen

1. Input query substance G.
   Initialization. Computation of $LSPL^G(\gamma_0, \gamma_1)$ $\forall$ distinct $\gamma_0, \gamma_1 \in Type(G)$, $Sim(G) = \emptyset$.
2. For $i = 1, \ldots, N$ do
   If $sumformula(G) = sumformula(G_i)$ then
   if $LSPL^G(\gamma_0, \gamma_1) = LSPL^{G_i}(\gamma_0, \gamma_1)$ $\forall$ distinct $\gamma_0, \gamma_1 \in Type(G)$ then $Sim(G) = Sim(G) \cup \{G_i\}$.
3. Output similarity list $Sim(G)$.

The function $sumformula(\cdot)$ in step 2 denotes the sum formula of a molecule. The output list contains all library compounds that are shortest path similar to the query compound. As this similarity notion is relational only, it does not yield similarity values.

4 Scaling Similarity

4.1 Voronoi Matrices

The notion of shortest path similarity can be scaled to be less discriminative but to allow a larger domain of comparisons. In particular, comparability of compounds that contain atoms of different atomar types becomes feasible. This is achieved by taking average values of all shortest path lengths that constitute second order lists of Graph Voronoi similarity.

The Voronoi distances between atoms of distinct types $\gamma_0, \gamma_1 \in Type(\Lambda)$ are given for any molecular graph G by $\delta_{Vor}(\gamma_0, \gamma_1, G) = \sum_{e \in LSPL^*(\gamma_0,\gamma_1),\, e \neq 0} \frac{1}{e}$ in case $\gamma_0, \gamma_1 \in \gamma(V)$ and $\gamma_0 \neq \gamma_1$. In all other cases the Voronoi distance is defined to be zero. The list $LSPL^*(\gamma_0, \gamma_1)$ is derived from the second order list $LSPL(\gamma_0, \gamma_1)$ by elimating all repetitions of entries that occur due to overlapping Voronoi regions sited at atoms of type γ_0.

For example, the second order list $LSPL(O, H)$ with respect to figure 1 has the five entries $3, 1, 3, 3, 3$, comp. the last table in sction 3.1. But the reduced second order list $LSPL^*(O, H)$ has only the four entries $3, 1, 3, 3$. One distance value 3 is elimated because the top center H-atom appears in two Voronoi regions for $\gamma_0 = O$.

4.2 Voronoi Matrix Similarity

The Voronoi distances can be arranged in a matrix called the Voronoi matrix of a molecular graph G. This matrix is given by

$$\Delta_{Vor}(G) = (\delta_{Vor}(\gamma_0, \gamma_1, G))_{\gamma_0,\gamma_1 \in Type(\Lambda)}.$$

These matrices are quadratic but unsymmetric in general. The distance between two molecular graphs can now be defined by the distance between two matrices. Here, we define the Voronoi distance between two molcules G_1, G_2 as

$$Dist_{Vor}(G_1, G_2) = ||\Delta_{Vor}(G_1) - \Delta_{Vor}(G_2)||_1.$$

The 1-norm $|| \cdot ||_1$ of a matrix is the maximum of column sums over absolute entries. Noteworthy, this is not a distance in the strict sense, but it serves as a measure denoting the distance between some query compound and each of the library compounds.

Instead of the 1-norm, the ∞-norm (maximum of row sums of absolute entries) and the Frobenius-norm (square root of sum of all squared entries) can be applied, see [12] for matrix norms. Two molecular graphs which are shortest path similar obviously have Voronoi distance zero, but the converse need not be true.

Voronoi matrix similarity is now defined in the obvious way. A molecular graph G_1 is more similar to a molecular graph G than another molecular garph G_2 if $Dist_{Vor}(G_1, G) \leq Dist_{Vor}(G_2, G)$.

4.3 Screening

Screening a library for Voronoi matrix similarity amounts to sorting the Voronoi matrix distances to the query compound. Since the query compound is comparable to all library compounds, all compounds are considered as "hits" and the crux lies in the distance values.

ScreenVMS

1. Input query substance G.
   Initialization. Computation of Voronoi matrix $D_{Vor}(G)$,
   $VS(G) = \Lambda$.
2. Sort $VS(G)$ according to increasing values of $Dist_{Vor}(G_i, G)$.
3. Output sorted Voronoi similarity list $VS(G)$.

This screening process allows various modifications such as cutting off all compounds whose distance from the query compound exceeds some externally given absolut threshold value. As an alternative, cutting off can be defined to retain only a small fraction of, for example, 5% of the library entries and sorting these.

5 Substructures

5.1 Distance Patterns

Instead of retrieving approximately similar compounds, all compounds may be searched for which contain the query compound. This problem is called substructure retrieval. Again, this retrieval problem is posed in an approximate sense. The essential means therefore is that of a distance pattern. A distance pattern is a shortest path matrix of a certain feature or descriptor. The concept is explained as follows.

The descriptors admitted for distance patterns are single atoms and specific functional groups. All occurrences of one functional group in a molecule must be free of vertex overlaps. For functional groups that appear more than once, the lengths of shortest paths between all occurrences are considered. A superstructure must have at least as many occurrences of the functional group as the query compound. A query compound with a sample descriptor and entries of the distance pattern are given in figure 4. A distance pattern can be represented by a complete graph with edge labels. For the query from figure 4 this graph is sketched in figure 5.

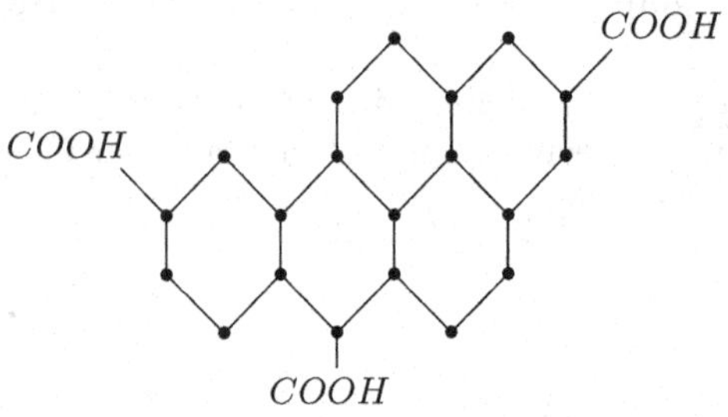

COOH

COOH

COOH

Fig. 4. Query structure with *COOH* as functional group. The distance pattern consists of the values 6, 8, 10.

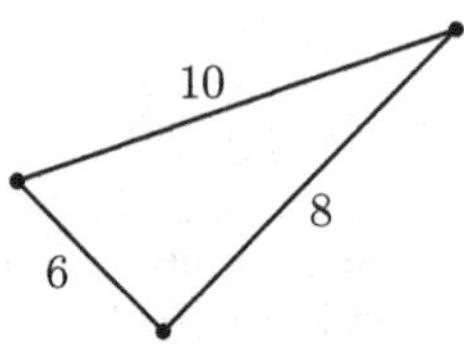

Fig. 5. Representation of the distance pattern from figure 4 by labeling the complete graph over three vertices.

5.2 Retrieval

Substructure retrieval searches for all superstructures whose distance patterns approximately cover the distance pattern of the query. Approximate coverage of a distance pattern is understood in the sense of the query distance pattern receiving the closest approximation by selections from the superstructure candidate. The only requirement is that the descriptor occurs as least as often in a superstructure candidate as in the query. A sample situation is given in figure 6.

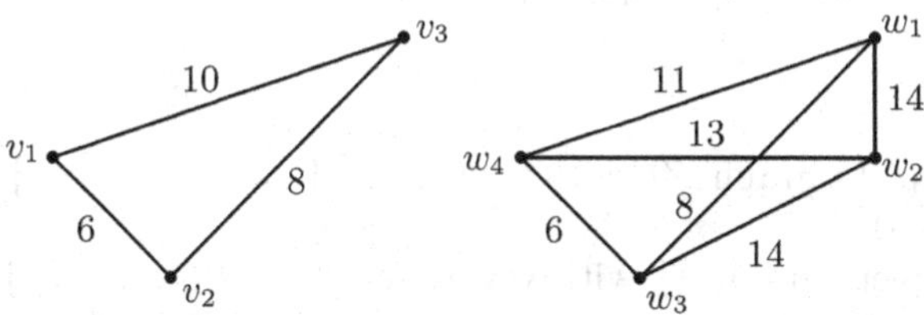

Fig. 6. Distance pattern of query (left) and distance pattern of a superstructure candidate (right). No selection of three vertices from the superstructure candidate is isomorphic to the query distance pattern. But the best selection of a complete subgraph with three vertices is considered as approximate cover of the original distance pattern.

Approximate coverage of distance patterns is formulated similar to classical best approximation problems. The best approximation of a complete labelled graph Z_1 with vertex set $V = \{v_1, \ldots, v_\nu\}$ and distance matrix $D(\cdot,\cdot)$ by some larger complete graph Z_2 with vertex set $W = \{w_1, \ldots, w_\mu\}$ and distance matrix $D'(\cdot,\cdot)$ is given as an invertible function $\varphi : V \to W$ solving the minimal matrix distance problem

$$\min_{\varphi:V\to W} ||(D(v_i, v_j)) - (D'(\varphi(v_i), \varphi(v_j)))||.$$

Feasible norms again include the Lebesgue norms and the Frobenius norm. The Lebesgue norms L_1 and L_∞ lead to the same values since distance patterns are symmetric matrices.

The best approximation for the situation from figure 6 is given by the function $\varphi(v_1) = w_4$, $\varphi(v_2) = w_3$, $\varphi(v_3) = w_1$ with vertex w_2 being unattained. The resulting minimum matrix distance is as follows

$$\begin{aligned} & ||\begin{pmatrix} 0 & 6 & 10 \\ 6 & 0 & 8 \\ 10 & 8 & 0 \end{pmatrix} - \begin{pmatrix} 0 & 6 & 11 \\ 6 & 0 & 8 \\ 11 & 8 & 0 \end{pmatrix}|| \\ = & ||\begin{pmatrix} 0 & 0 & -1 \\ 0 & 0 & 0 \\ -1 & 0 & 0 \end{pmatrix}|| = \begin{cases} 1, & \text{if } ||\cdot|| = ||\cdot||_1 \text{ or } ||\cdot|| = ||\cdot||_\infty \\ \sqrt{2}, & \text{if } ||\cdot|| = ||\cdot||_F. \end{cases} \end{aligned}$$

The best approximation is the same in this example irrespective of 1-norm, the ∞-norm or the Frobenius norm being used.

The best approximation problem can be solved by enumeration which is finite here. A faster, approximate computing scheme makes approximate coverage decisions by distance lists instead of distance matrices. The idea is a greedy procedure that traverses the smaller graph vertex-wise. The procedure assigns a minimum distance vertex from the yet unused part of the larger graph to each vertex of the yet unused part of the smaller graph. Distances are computed as vector distances between distance lists. Iterative selections are performed until each vertex of the smaller graph has received an assignment. This results in the subsequent procedure for which details are given in [14].

AppCov

1. Input complete graph Z_1 with vertex set $V = \{v_1, \ldots, v_\nu\}$ and distance matrix $D(\cdot,\cdot)$
   complete graph Z_2 with vertex set $W = \{w_1, \ldots, w_\mu\}$ and distance matrix $D'(\cdot,\cdot)$, $\nu \leq \mu$.
   Initialization. $A = V$, $B = W$.
2. While $A \neq \emptyset$ do
   a) Computation of
   $$w(v) = argmin_{w \in B} \delta(v, w)$$
   for all $v \in A$ with $\delta(v, w) = \min_{pr_\nu} ||distlist(v) - pr_\nu(distlist(w))||$.
   b) Selection of $v_0 = argmin_{v \in A} \delta(v, w(v))$.
   c) $\varphi(v_0) = w(v_0)$.
   d) $A = A - \{v_0\}$.
   e) $B = B - \{\varphi(v_0)\}$.
3. Output $\varphi(\cdot)$ and $||(D(v_i, v_j)) - (D'(\varphi(v_i), \varphi(v_j)))||$.

All distance lists in step 2 are increasingly sorted and the projections pr_ν denote selections of ν out of μ values. Once distance lists are sorted, all selections needed to be considered are fixed length intervals of consecutive values. Only a linear number of these, namely $\mu - \nu + 1$ exist. A rough upper bound on the run time of algorithm **AppCov** is $O(\nu^2 \mu(\mu - \nu + 1))$. The average time complexity can be expected to be much lower if many of the recomputations in step 2(a) can be avoided.

The foregoing algorithm is not ensured to find the best approximation even if the best approximation is unique with minimum matrix distance zero. But the best approximation is found if there is a unique choice with error zero in each iteration of the second step.

5.3 Approximate Screening

Screening the library $\Lambda = (G_i)_{i=1}^N$ with distance patterns $(D_i(\cdot,\cdot))_{i=1}^N$ for superstructures can now be accomplished by the following procedure.

ScreenApproxSuper

1. Input query substance G and descriptor.
   Initialization. Computation of $\nu = descoccur(G)$,
   $AS(G) = \emptyset$.
2. For $i = 1, \ldots, N$ do
   If $descoccur(G_i) \geq \nu$ then $AS(G) = AS(G) \cup \{G_i\}$.
   Sort $AS(G)$ by increasing distances $||(D(v_k, v_l)) - (D_i(\varphi_i(v_k), \varphi(v_l)))||$.
3. Output sorted superstructure list $AS(G)$.

The functions φ_i are the subisomorphism found by algorithm **AppCov** when applied to the distance pattern of G and to the distance pattern of any qualifying compound G_i as larger complete graph. The size of the output of the algorithm **ScreenApproxSuper** can be reduced by setting a threshold value that limits the matrix distance between query and library compounds.

The given approach can extended to allow for multiple descriptors in a straightforward way. First, a list is generated that contains all compounds in which each of the descriptors occurs at least as often as in the query pattern. This list is computed by sequentially thinning out lists according to the descriptor occurrence criterion. Then, the final list is sorted according to increasing sums of matrix distances with the sums being taken over all descriptors. When considering several instead of a single distance pattern, some descriptors are allowed to occur only once or even not at all in the query.

References

[1] Advanced Chemistry Development Inc. ChemFolder. Toronto, www.acdlabs.com

[2] Artymiuk, P.J. et al.: Similarity searching in data bases of three-dimensional molecules and macromolecules. Journal of Chemical Information and Computer Sciences **32** (1992) 617-630

[3] Beilstein Informationssysteme GmbH, now MDL Information Systems GmbH, Frankfurt, www.beilstein.com

[4] CambridgeSoft corporation, Cambridge, MA. ChemOffice. www.camsoft.com.

[5] Devillers, J., Balaban, A.T.: Topological indices and related descriptors in QSAR and QSPR. Gordon and Breach, London (1999)

[6] Dury, L., Latour, T., Leberte, L., Barberis, F., Vercauteren, D.P.: A new graph descriptor for molecules containing cycles. Journal of Chemical Information and Computer Sciences **41** (2001) 1437-1445

[7] Erwig, M.: The graph Voronoi diagram with applications. Networks **36** (2000) 156-163

[8] Faulon, J.-L.: Isomorphism, automorphism partitioning, and canonical labeling can be solved in polynomial-time for molecular graphs. Journal of Chemical Information and Computer Sciences **38** (1998) 432-444

[9] Fligner, M., Verducci, J., Bjoraker, J., Blower, P.: A new association coefficient of molecular dissimilarity. (2001) cisrg.shef.ac.uk/ shef2001/talks/blower.pdf.

[10] Gillet, V.J., Wild, D.J., Willett, P., Bradshaw, J.: Similarity and dissimilarity methods for processing chemical structure databases. The Computer Journal **41** (1998) 547-558

[11] Goldman, D., Istrail, S., Lancia, G., Piccolboni, A., Walenz, B.: Algorithmic strategies in combinatorial chemistry. Proceedings of Symposium on Discrete Algorithms SODA (2000) 275-284

[12] Golub, G.H., van Loan, C.F.: Matrix computations. John Hopkins University Press, 4th printing, Baltimore (1985)

[13] Ivanovic, O., Klein, D.J.: Computing Wiener-type indices for virtual combinatorial libraries generated from heteroatom-containing building blocks. Journal of Chemical Information and Computer Sciences **42** (2002) 8-22

[14] Kämpke, T.: Distance patterns in structural similarity. (2003) submitted

[15] Kämpke, T., Schaal, M.: Distributed generation of fastest paths. Proceedings of International Conference on Parallel and Distributed Computing and Systems PDCS '98, Las Vegas (1998) 172-177

[16] Lobanov, V.S., Agrafiotis, D.K.: Stochastic similarity selections from large combinatorial libraries. Journal of Chemical Information and Computer Sciences **40** (2000) 460-470

[17] Molecular Networks GmbH. Carol. Erlangen, www.mol-net.de.

[18] Richter, M.M.: The knowledge contained in similarity measures. www.cbr-org.web/documents/Richtericcbr95remarks.html.

[19] Rouvray, D.H.: The topological matrix in quantum chemistry. In: Balaban, A.T. (ed.): Chemical applications of graph theory. Academic Press, New York (1976) 175-221.

[20] Todeschini, R., Consonni, V.: Handbook of molecular descriptors. Wiley-VCH, Weinheim (2000).

Engineers Don't Search

Benno Stein

Paderborn University
Department of Computer Science
D-33095 Paderborn, Germany
stein@upb.de

Abstract. This paper is on the automation of knowledge-intensive tasks in engineering domains; here, the term "task" relates to analysis and synthesis tasks, such as diagnosis and design problems.

In the field of Artificial Intelligence there is a long tradition in automated problem solving of knowledge-intensive tasks, and, especially in the early stages, the search paradigm dictated many approaches. Later, in the modern period, the hopelessness in view of intractable search spaces along with a better problem understanding led to the development of more adequate problem solving techniques.

However, search still constitutes an indispensable part in computer-based diagnosis and design problem solving—albeit human problem solvers often gets by without: "Engineers don´t search" is my hardly ever exaggerated observation from various relevant projects, and I tried to learn lessons from this observation. This paper presents two case studies.

1. Diagnosis problem solving by model compilation. It follows the motto: "Spend search in model construction rather than in model processing."
2. Design problem solving by functional abstraction. It follows the motto: "Construct a poor solution with little search, which then must be repaired."

On second sight it becomes apparent that the success of both mottos is a consequence of untwining logic-oriented reasoning (in the form of search and deduction) and approximation-oriented reasoning (in the form of simulation).

Keywords: Model Construction, Search, Diagnosis, Design Automation

1 Automating Knowledge-Intensive Tasks

"How can knowledge-intensive tasks such as the diagnosis or the design of complex technical systems be solved using a computer?"

A commonly accepted answer to this question is: "By operationalizing expert knowledge!" And in this sense, the next subsection is a hymn to the simple but powerful, associative models in automated problem solving. Engineers don't search,[1] and computer programs that operationalize engineer (expert) knowledge have been proven successful in various complex problem solving tasks.

[1] Which also means: "Experts don't search", or "need less search" (during problem solving).

W. Lenski (Ed.): Logic versus Approximation, LNCS 3075, pp. 120–137, 2004.

A second, also commonly accepted answer to the above posed question is: "By means of search!" This answer reflects the way of thinking of the modern AI pragmatist, who believes in deep models and the coupling of search and simulation. Deep models, or, models that rely on "first principles" have been considered the worthy successor of the simple associative models [6, 9]; they opened the age of the so-called Second Generation Expert Systems [8, 37]. In this sense, Subsection 1.2 formulates diagnosis and design problems as instances of particular search-plus-simulation problems.

Though the search-plus-simulation paradigm can be identified behind state-of-the-art problem solving methodologies [2, 14], many systems deployed in the real world are realized according to simpler associative paradigms (cf. [4, 5, 24, 28, 32, 34], to mention only a few). As a source for this discrepancy we discover the following connection: The coupling of search and simulation is willingly used to make up for missing problem solving knowledge but, as a "side effect", often leads to intractable problems.

Drawing the conclusion "knowledge over search" is obvious on the one hand, but too simple on the other: Among others, the question remains what can be done if the resource "knowledge" is not available or cannot be elicited, or is too expensive, or must tediously be experienced? Clearly, expert knowledge cannot be cooked up—but we learn from human problem solvers where to spend search effort deliberately in order to gain the maximum impact for automated problem solving. The paper in hand gives two such examples: In Subsection 1.3 we introduce the principles of "model compilation" and "functional abstraction" to address behavior-based diagnosis and design problems. These principles are fairly different and specialized when compared to each other; interestingly, common to both is that they develop from the search-plus-simulation paradigm by untwining the roles of search and simulation. In this way they form a synthesis of the aforementioned paradigms.

The Sections 2 and 3 of this paper outline two case studies from the field of fluidic engineering, which illustrate how the proposed principles are put to work.

1.1 Thesis: Knowledge Is Power[2]

Human problem solving expertise is highly effective but of heuristic nature; moreover, it is hard to elicit but rather easy to process [21]. E. g., a simple but approved formalization of diagnosis knowledge are associative connections:

$$obs_1 \wedge \ldots \wedge obs_k \rightarrow d,$$

where the obs_i and d denote certain observations and a diagnosis respectively. Likewise, successful implementations of design algorithms don´t search in a gigantic space of behavior models but operate in a well defined structure space instead, which is spanned by compositional (left) and taxonomic relations (right):

$$c \rightarrow c_1 \wedge \ldots \wedge c_k \qquad\qquad c \rightarrow c_1 \vee \ldots \vee c_k,$$

where the c_i denote components, i. e., the associations describe a decomposition hierarchy in the form of an And-Or-graph. Another class of design algorithms employ the

[2] This famous phrase is often attributed to Edward A. Feigenbaum, though he did not originate the saying.

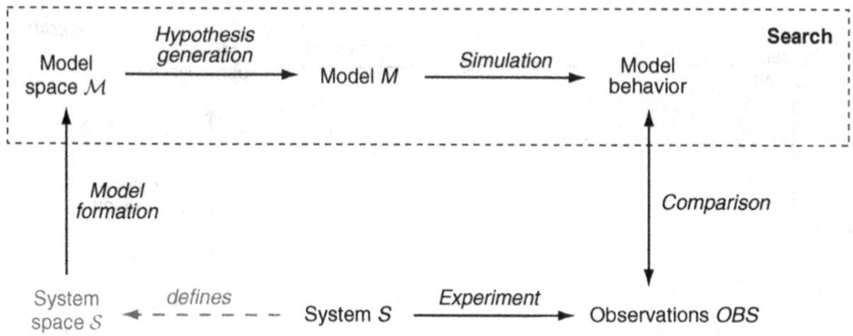

Fig. 1. A generic scheme of model-based diagnosis: Given is the interesting system S, which defines a space $\mathcal{S}$ of faulty systems, and a set of observations OBS. On a computer, $\mathcal{S}$ is represented as a model space, $\mathcal{M}$, wherein a model M^* is searched whose simulation complies with OBS.

case-based reasoning paradigm retrieve-and-adapt, an advancement of the classical AI paradigm generate-and-test [22, 30, 31]:

$$SIM(D_1, D_2) \rightarrow USABILITY(M_1, D_2),$$

which states that the known solution M_1 for a demand set D_1 can be used (adapted) to satisfy a demand set D_2, if D_1 and D_2 are similar.

1.2 Antithesis: Search Does All the Job

Preliminaries. Let S be a system. In accordance with Minsky we call M a model of S, if M can be used to answer questions about S [25]. M may establish a structural, a functional, an associative, or a behavioral model. In this paper the focus is on behavioral models, which give us answers to questions about a system's behavior.

A search problem is characterized by a search space consisting of states and operators. The states are possible complete or partial solutions to the search problem, the operators define the transformation from one state into another. Here, in connection with behavior-based diagnosis and design problems, the search space actually is a model space. It is denoted by $\mathcal{M}$. The model space is only defined implicitly; it comprises all models that could be visited during search.

Diagnosis Problem Solving. Starting point of a diagnosis problem is a system S along with as set of observations OBS. The observations are called symptoms if they do not coincide with the expected behavior of S. Performing diagnosis means to explain symptoms in terms of misbehaving components, that is, to identify a system S^* in a space $\mathcal{S}$ of faulty systems that will exhibit OBS. A model-based diagnosis algorithm performs this search in a model space $\mathcal{M}$, which contains—at the desired level of granularity—models that correspond to faulty systems in $\mathcal{S}$. The objective is to identify a model $M^* \in \mathcal{M}$ whose simulation produces a behavior that complies with OBS. Figure 1 illustrates the connections.

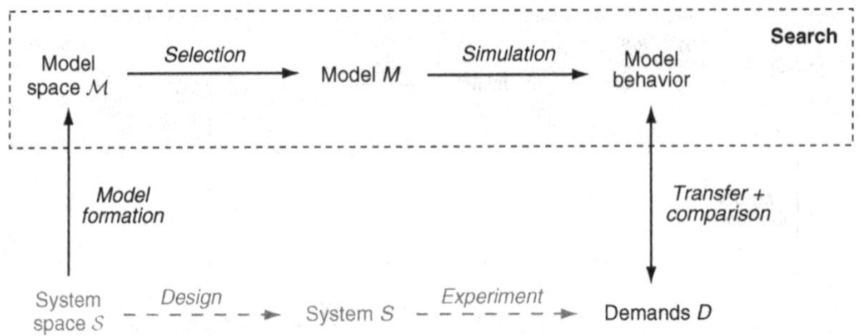

Fig. 2. An generic scheme of design problem solving: Given is a space $\mathcal{S}$ of possible design solutions and a set of demands D. On a computer, $\mathcal{S}$ is represented as a model space, $\mathcal{M}$, wherein a model M^* is searched whose behavior fulfills D.

Several model-based diagnosis approaches, such as the GDE, GDE+, or Sherlock base on such a simulation cycle. Their search in $\mathcal{M}$ is highly informed since they exploit the underlying device topology for hypotheses generation.[3] Adopting the notation of Reiter, model-based diagnosis can be formalized as follows [29]:

$$\alpha_\Delta = \mathit{SD} \wedge \mathit{OBS} \wedge \{AB(c) \mid c \in \Delta\} \wedge \{\neg AB(c) \mid c \in \mathit{COMPS} \setminus \Delta\},$$

where SD is a logic-based formulation of $\mathcal{M}$, COMPS denotes a set of symbols that represent the components of the system S, $\Delta \subseteq \mathit{COMPS}$ denotes the broken components, and AB is a special predicate that indicates whether or not a component is working abnormally. Stipulating Occam's razor, a diagnosis algorithm determines a diagnosis Δ as the solution of the following optimization problem:

$$\Delta = \text{argmin}_i \Big(|\Delta| = i \;\; \wedge \;\; \alpha_\Delta \text{ is satisfiable} \Big)$$

Design Problem Solving. Starting point of a design problem is a space $\mathcal{S}$ of possible design solutions along with a set D of demands. Solving a design problem means to determine a system $S^* \in \mathcal{S}$ that fulfills D. Typically, S^* is not found by experimenting in the real world but by operationalizing a virtual search process after having mapped the system space, $\mathcal{S}$, onto a model space, $\mathcal{M}$. It is the job of a design algorithm to efficiently find a model $M^* \in \mathcal{M}$ whose simulation produces a behavior that complies with D and which optimizes a possible goal criterion. Figure 2 illustrates the connections.

Compared to the previous diagnosis scheme, the model space of a design problem is usually orders of magnitude bigger. This is also reflected by the following formalization:

$$\alpha_{CPT} = \underbrace{\mathit{SD} \wedge D \wedge \mathit{COMPS}}_{\text{configuration}} \underbrace{\wedge\, \mathit{PARAM}}_{\text{+ parameterization}} \underbrace{\wedge\, \mathit{TOP}}_{\text{+ structure finding}} ,$$

[3] Model-based diagnosis approaches can be further characterized in the way simulation is controlled, fault models are employed, dependencies are recorded, measurement points are chosen, or failure probabilities are utilized [17, 11, 12, 42].

where SD is a logic-based formulation of the behavior of all components in $\mathcal{M}$, $COMPS$ denotes the actually selected components, $PARAMS$ their parameterization, and TOP defines the topology, i. e., how the selected components are connected to each other. The complexity of a design problem depends on the degrees of freedom in the search process: Within a configuration problem merely $COMPS$ is to be determined, whereas in a behavior-based design problem the components need to be parameterized and yet the system structure is to be found such that α_{CPT} is satisfiable.

The outlined diagnosis and design schemes are inviting: Giving a mapping from systems $\mathcal{S}$ to models $\mathcal{M}$—which can be stated straightforwardly in engineering domains—the related analysis or synthesis problem can be solved by the search-plus-simulation paradigm. As already mentioned at the outset, many successful implementations of diagnosis and design systems do not follow this paradigm. They contain an explicit representation of an engineer's problem solving knowledge instead, say, his or her model of expertise. A problem solver that has such knowledge-based models at its disposal spends little effort in search—a fact which makes these models appearing superior to the deep models used in the search-plus-simulation paradigm. On the other hand, several arguments speak for the latter; a compelling one has to do with knowledge acquisition: In many situations it is not feasible for technical or economical reasons to acquire the necessary problem solving knowledge to operationalize tailored models of expertise.[4] Remarkably, de Kleer actually concludes: " Knowledge isn't power. Knowledge is evil." [10].

1.3 A Synthesis: Untwine Search and Simulation

The purpose of this subsection is twofold: It annotates problems of the search-plus-simulation paradigm, and it introduces two advancements of this principle: model compilation and functional abstraction. Here, the former is applied to diagnosis problem solving, while the latter is used to tackle a design problem.[5] Within both principles search as well as simulation still play central roles. However, compared to the search-plus-simulation paradigm, the simulation step is no longer integral part of the search cycle, say, search (logic-oriented reasoning) and simulation (approximation-oriented reasoning) are untwined.

Model Compilation. The search-plus-simulation paradigm in model-based diagnosis enables one to analyze a system for which no diagnosis experience is available or which is operated under new conditions [17]. On the other hand, for complex technical systems model-based diagnosis needs excellent simulation capabilities, because the goal driven reasoning process requires inverse simulation runs (from observations back to causes) to efficiently cover all symptoms [14, 11]. Still more problematic are the following limitations:

[4] Other advantages bound up with this paradigm are: the possibility to explain, to verify, or to document a reasoning process, the possibility to reuse the same models in different contexts, the extendibility to new device topologies, or the independence of human experts.

[5] This correspondence is not obligatory; in [39] a configuration tool of a large telecommunication manufacturer is described, wherein model compilation provides a key technology.

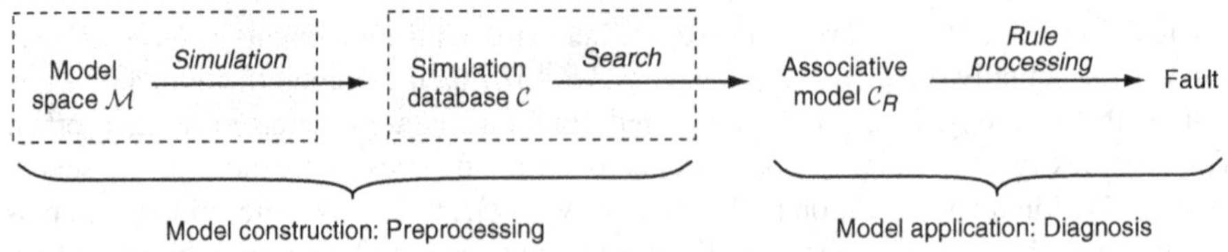

Fig. 3. Model compilation untwines logic-oriented and approximation-oriented reasoning: Simulating the model of a system in various fault modes yields a simulation data base $\mathcal{C}$ from which a rule-based model $\mathcal{C}_R$ is constructed.

1. In domains with continuous quantities the classification of values as symptoms is ambiguous [23].
2. Long interaction paths between variables result in large conflict sets.
3. Many technical systems have a feedback structure; i. e., cause-effect chains, which are the basis for an assumption-based reasoning process, cannot be easily stated.

A promising strategy in this situation is the compilation of an associative model from the given model of first principles, which is achieved as follows: By simulating the model of first principles in various fault modes and over its typical input range a simulation database $\mathcal{C}$ is built up. Within a subsequent search step, a rule-based model $\mathcal{C}_R$ is constructed from $\mathcal{C}$, where long cause-effect chains are replaced with weighted associations and which is optimized for a classification of the fault modes. In this way the simulation and the search step form a preprocessing phase, which is separated from the phase of model application, i. e., the diagnosis phase. Figure 3 illustrates the principle.

Compiled models have a small computational footprint. As well as that, model compilation breaks feedback structures, and, under the assumption that all observations have already been made, an optimum fault isolation strategy can be developed [41]. Section 2 outlines a model compilation application in the fluidic engineering domain.

Functional Abstraction. The search-plus-simulation paradigm has also been suggested as a fundamental problem solving strategy for design tasks.[6] While the role of simulation is like in the diagnosis task above, namely, analyzing a model's behavior, does the reasoning situation raise another difficulty: Applying just search-plus-simulation renders real-world design tasks intractable, because of the mere size of the related model space $\mathcal{M}$. As already indicated, the lack of problem solving knowledge (here in the form of design rules) forces one to resort to the search-plus-simulation paradigm. Again, model compilation could be applied to identify underlying design rules, but, this is tractable only for medium-sized configuration problems [39]. Moreover, the complexity problem, which is caused by the size of $\mathcal{M}$, is not eased but only shifted to a preprocessing phase.

If search cannot be avoided, one should at least ensure that search effort is spent deliberately. In this situation we learn from the problem solving behavior of engineers:

---

[6] Gero, for example, proposes a cycle that consists of the steps synthesis, analysis, and evaluation. Sinha et al. present a framework to implement simulation-based design processes for mechatronic systems [27, 35].

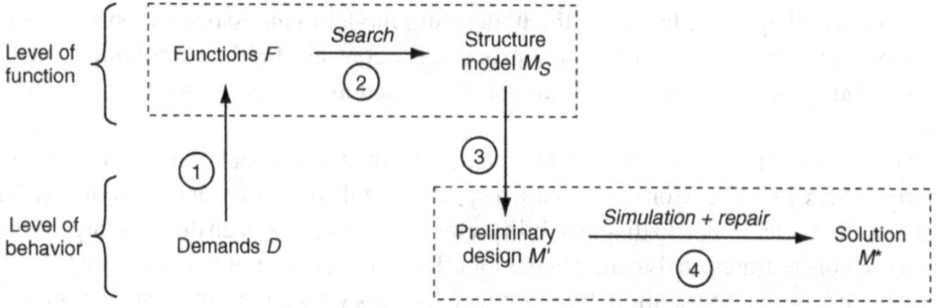

Fig. 4. The paradigm of functional abstraction applied to design problem solving. Observe that logic-oriented reasoning (search) has been decoupled from approximation-oriented reasoning (simulation + repair): The former is used to find a structure model M_S, the latter is used to repair a suboptimum raw design.

1. Engineers solve a design problem rather at the level of function than at the level of behavior, accepting to miss the optimum.
2. Engineers rather adapt a suboptimum solution than trying to develop a solution from scratch, accepting to miss the optimum.
3. Engineers can formulate repair and adaptation knowledge easier than a synthesis theory.

Putting together these observations one obtains the paradigm "Design by Functional Abstraction", which is illustrated in Figure 4.[7] Put it overstated, the paradigm says: At first, we construct a poor solution of a design problem, which then must be repaired.

Key idea of design by functional abstraction is to construct candidate solutions within a very simplified design space, which typically is some structure model space. A candidate solution, M_S, is transformed into a preliminary raw design, M', by *locally* attaching behavior model parts to M_S. The hope is that M' can be repaired with reasonable effort, yielding an acceptable design M^*. Design by functional abstraction makes heuristic simplifications at two places: The original demand set, D, is simplified toward a functional specification F (Step 1), and, M_S is transformed locally into M' (Step 3). Section 3 presents an application of this paradigm.

2 Case Study I. Diagnosis Problem Solving by Model Compilation

The fault detection performance of a diagnosis system depends on the adequateness of the underlying model. Model compilation is one paradigm for constructing adequate models; the model-based diagnosis paradigm, either with or without fault models, provides another. Under the latter, the cycle of simulation and candidate discrimination is executed at runtime, while under the compilation paradigm it is anticipated in a preprocessing phase (see Figures 1 and 3). Reasoning with compiled diagnosis models is similar

[7] The first three steps of this method resemble syntax and semantics (the horseshoe principle) of the problem solving method "Heuristic Classification", which became popular as the diagnosis approach underlying MYCIN [7].

to associative diagnosis; however, the underlying model in an associative system is the result of a substantial model formation process. By contrast, model compilation pursues a data mining strategy and aims at an automatic acquisition of associative knowledge [36].

The idea to derive associative knowledge from deep models has been proposed among others in [37]. Moreover, with respect to fault detection and isolation (FDI), measurement selection, and diagnosability a lot of research has been done. A large part of this work concentrates on dynamic behavior effects, which are not covered here [15, 26]. Nevertheless, since the compilation concept focuses on search space and knowledge identification aspects it can be adapted to existing FDI approaches as well.

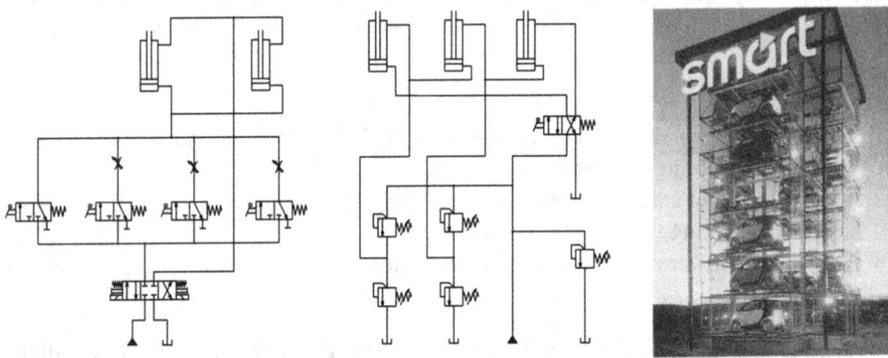

Fig. 5. Diagrams of two medium-sized hydraulic circuits and a photo of the hydraulically operated Smart Tower.

2.1 Hydraulic Systems and Their Components

In this section, as well as in Section 3, the field of hydraulic engineering serves us as application domain. Hydrostatic drives provide advantageous dynamic properties and represent a major driving concept for industrial applications. They consist of several types of hydraulic building blocks: Cylinders, which transform hydraulic energy into mechanical energy, various forms of valves, which control flow and pressure of the hydraulic medium, and service components such as pumps, tanks, and pipes, which provide and distribute the necessary pressure p and flow Q. Figure 5 (left-hand side) shows two medium-sized examples of circuits we are dealing with.

Component Faults and Fault Models. A prerequisite for applying model compilation for diagnosis purposes is that components are defined with respect to both their normal and their faulty behavior. Below, such a fault model is stated exemplary for the check valve. Typical check valve faults include jamming, leaking, or a broken spring. These faults affect the resistance characteristic of the valve in first place (cf. Table 1).

Other fault models relate to slipping cylinders due to interior or exterior leaking, incorrect clearance or sticking throttle valves, directional valves with defect solenoid or

Table 1. Resistance law of a working and a faulty check valve operating in its control range where $\Delta p > p_O$. The deviation coefficient $\varepsilon_{\text{valve}}$ is a state quantity, which is modeled as a continuous random variable.

Normal resistance behavior	Faulty resistance behavior
$R = \dfrac{m^2 \cdot \Delta p}{(\Delta p - p_O)^2}$	$R = \dfrac{m^2 \cdot \Delta p}{(\Delta p - p_O \cdot (1 + \varepsilon_{\text{valve}}))^2}$

contaminated lands, and pumps showing a decrease in performance. For all fault models, a deviation coefficient ε is modeled as a continuous random variable which defines the distribution of the fault seriousness.

2.2 Construction of a Compiled Model

We construct a compiled model for a system S in five steps. Within the first step a simulation data base $\mathcal{C}$ is built, which is then successively abstracted towards a real-valued symptom data base $\mathcal{C}_\Delta$, a symbolic interval data base $\mathcal{C}_I$, an observer data base $\mathcal{C}_O$, and, finally, a rule set $\mathcal{C}_R$, which represents the heuristic diagnosis model.[8]

Simulation. Behavior models of hydraulic systems are hybrid discrete-event/continuous-time models [3]. The trajectories of the state variables can be considered as piecewise continuous segments, called phases. The discrete state variables such as valve positions, relays, and switches are constant within a phase, and in between the phases one or more of them changes its value, leading to another mode of the system. The continuous variables z_i such as pressures, flows, velocities, or positions are the target of our learning process; they form the set Z. The phase-specific, quasi-stationary values of the variables in Z are in the role of symptoms, since abrupt faults can cause their significant change.

Let S be a system, let D be a set denoting the interesting component faults in S, and let $\mathcal{M}$ be the related space of models. I. e., $\mathcal{M}$ includes the interesting faulty models with respect to D as well as the correct model of S. The result of the simulation step is a data base $\mathcal{C}$, which contains samples of the vector of state trajectories drawn during the simulation of the models in $\mathcal{M}$.

Symptom Identification. For each fault simulation vector in $\mathcal{C}$ the deviations of its state variables to the related faultless simulation vector is computed. The computation is based on a special operator "$\ominus$", which distinguishes between effort variables and flow variables. The former are undirected, and a difference between two values of this type is computed straightforwardly. The latter contain directional information, and their difference computation distinguishes several cases. Result of this step is the symptom data base of $\ominus$-deviations, $\mathcal{C}_\Delta$.

Interval Formation. The symptom vectors in $\mathcal{C}_\Delta$ are generalized by mapping for each $z \in Z$ the deviations $\delta_z^{(1)}, \ldots, \delta_z^{(|\mathcal{C}|)}$, $\delta_z^{(i)} \in \mathbf{R}$, onto p intervals $\mathbf{I}_z^{(1)}, \ldots, \mathbf{I}_z^{(p)}$, $\mathbf{I}_z^{(j)} \subset \mathbf{R}$, with $\bigcup_j \mathbf{I}_z^{(j)} = \mathbf{R}$. This is an optimization task where, on the one hand, the loss of

[8] A detailed description of the compilation procedure can be found in [40].

discrimination information is to be kept minimum (the larger p the better), while on the other hand, constraints of measuring devices are to be obeyed (the smaller p the better). Observe that in this abstraction step the domain of real numbers is replaced by a propositional-logical representation: For each state variable $z \in Z$ a new domain I_z of interval names is introduced, which map in a one-to-one manner onto the real-valued intervals. The symbolic interval database that develops from $\mathcal{C}_\Delta$ by interval formation is denoted by $\mathcal{C}_I$.

Measurement Selection. By means of simulation, values are computed for all variables in Z. In fact, restricted to a handful of measuring devices or sensors, only a small subset $O \subset Z$ can be observed at the system. Measurement selection means to determine the most informative variables in Z—or, speaking technically, to place a set of $|O|$ observers such that as many faults as possible can be classified. O is determined by analyzing for each phase and for each variable $z \in Z$ the correlations between the symbolic intervals I_z and the set of component faults D. Our analysis generalizes the idea of hypothetical measurements, which goes back on the work of Forbus and de Kleer. Let $O \subset Z$ be the set of selected observers according to this analysis; the database that emerges from the symbolic interval database $\mathcal{C}_I$ by eliminating all variables in $Z \setminus O$ is called observer database $\mathcal{C}_O$; it is much smaller than $\mathcal{C}_I$. However, its number of elements is unchanged, i. e., $|\mathcal{C}_O| = |\mathcal{C}|$.

Rule Generation. Within the rule generation step reliable diagnosis rules are extracted from $\mathcal{C}_O$. The rules have a propositional-logical semantics and are of the form

$$\mathbf{r} = \iota_1 \wedge \ldots \wedge \iota_k \rightarrow d,$$

where the ι_i denote interval names and d denotes a diagnosis. The semantics of $\mathbf{r}$ is defined by means of two truth assignment functions, $\alpha : \bigcup I_z \rightarrow \{0,1\}$ and $\beta : D \rightarrow \{0,1\}$. If $\mathbf{I}$ is the real-valued interval associated with the interval name ι and if δ is a symptom observed at S, then α and β are defined as follows:

$$\alpha(\iota) = \begin{cases} 1 \Leftrightarrow \delta \in \mathbf{I} \\ 0 \text{ otherwise.} \end{cases} \qquad \beta(d) = \begin{cases} 1 \Leftrightarrow \text{ the fault in } S \text{ is } d. \\ 0 \text{ otherwise.} \end{cases}$$

Note that the inference direction of these rules is reverse to the cause-effect computations when simulating a behavior model: We now ask for symptoms and deduce faults, and—as opposed to the simulation situation—this inference process must not be definite. To cope with this form of uncertainty each rule $\mathbf{r}$ is characterized by (1) its confidence $c(\mathbf{r})$, which rates the logical quality of the implication, and (2) its relative frequency, called "support" in the association rule jargon [1]. The rule generation step is realized with data mining methods and yields the desired rule model $\mathcal{C}_R$.

2.3 Experimental Analysis: Diagnosis with DÉJÀVU

Diagnosing the system S means to process the rule model $\mathcal{C}_R$ subject to the observed symptoms. For the operational semantics of the rules' confidence and support values we employ the formula below, which computes for each fault $d \in D$ its confidence in

"$\beta(d) = 1$", given a rule model $\mathcal{C}_R$ and a truth assignment α. The formula combines the highest achieved confidence with the average confidence of all matching rules:

$$c(\text{"}\beta(d) = 1\text{"}) = c(\mathbf{r}^*) + \big(1 - c(\mathbf{r}^*)\big) \cdot \frac{1}{|\mathcal{R}|} \sum_{\mathbf{r} \in \mathcal{R}} c(\mathbf{r}),$$

where $\mathcal{R} \subset \mathcal{C}_R$ denotes the matching rules with conclusion d, and $\mathbf{r}^* \in \mathcal{R}$ denotes a rule of maximum confidence.

The outlined model construction process as well as the rule inference have been operationalized within the diagnosis program DÉJÀVU [20]. For simulation purposes, DÉJÀVU employs the FLUIDSIM simulation engine. The approach has been applied to several medium-sized hydraulic circuits (about 20-40 components) with promising results. The table in Figure 6 shows the diagnosis performance depending on the number of observers $|O|$; basis were more than 2000 variations of $|D| \approx 15$ different component faults. The results were achieved with automatically constructed rule models $\mathcal{C}_R$ that have not been manually revised. The right-hand side of Figure 6 shows the average number of rules in $\mathcal{C}_R$ as a function of $|O|$.

Diagnosis accuracy	Number of observers $\lvert O\rvert$ 1	2	3	4	5	6
"exact"	0.40	0.55	0.53	0.52	0.52	0.53
"1 in 3"	0.51	0.72	0.81	0.88	0.92	0.96

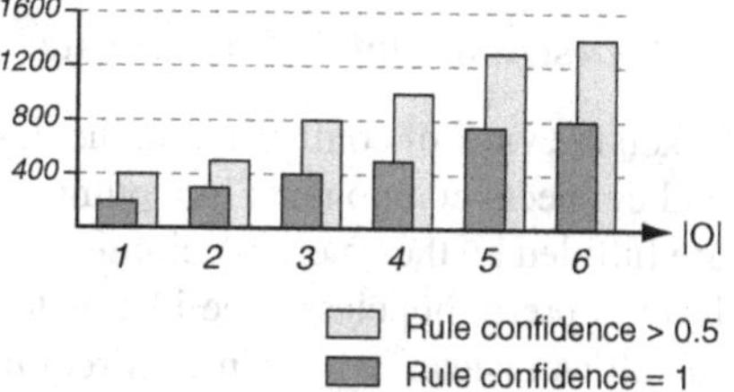

Fig. 6. The table shows the fraction of classified faults depending on the observer number $|O|$; "exact" stands for a unique fault prediction, "1 in 3" indicates a multiple prediction of two or three components including the faulty one. The bar graph on the right shows the number of generated rules, $|\mathcal{C}_R|$, as a function of $|O|$. Dark bars indicate rules with a confidence value of 1, light bars stand for confidence values greater than 0.5.

3 Case Study II. Design Problem Solving by Functional Abstraction

Even for an experienced engineer, the design of a fluidic system is a complex and time-consuming task, that, at the moment, cannot be automated completely. The effort for acquiring the necessary design knowledge exceeds by far the expected payback, and, moreover, the synthesis search space is extremely large and hardly to control—despite the use of knowledge-based techniques.

Two possibilities to counter this situations are "competence partitioning" and "expert critiquing". The idea of competence partitioning is to separate the creative parts of a design process from the routine jobs, and to provide a high level of automation regarding the latter [38]. Expert critiquing, on the other hand, employs expert system technology to assist the human expert rather than to automate a design problem in its entirety [18, 13].

In this respect, design by functional abstraction can be regarded as a particular expert critiquing representative.

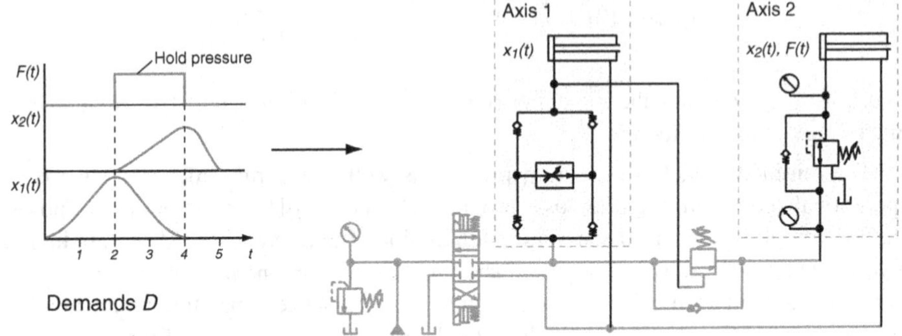

Fig. 7. Hydraulic design means to translate a demand description (left) to a circuit model.

3.1 Design in Fluidic Engineering

Taken the view of configuration, the designer of a fluidic system selects, parameterizes, and connects components like pumps, valves, and cylinders such that the demands D are fulfilled by the emerging circuit.[9] Solving a fluidic design problem at the component level is pretty hopeless. The idea is to perform a configuration process at the level of functions instead, which in turn requires that fluidic functions possess constructional equivalents that can be treated in a building-block-manner. In the fluidic engineering domain this requirement is fairly good fulfilled; the respective building blocks are called "fluidic axes".

Figure 7 shows a demand description D (left) and a design solution M^* in the form of a circuit diagram that fulfills D. The circuit consists of two hydraulic axes that are coupled by a sequential coupling. To automate this design process, so to speak, to automate the mapping $D \longrightarrow M^*$, we apply the paradigm of functional abstraction (cf. Figure 8 and recall Figure 4):

1. The demand specification, D, is abstracted towards a functional specification, F.
2. At this functional level a structure model M_S according to the coupling of the fluidic functions in F is generated.
3. M_S is completed towards a tentative behavior model M' by plugging together locally optimized fluidic axes; here, this step is realized by case-based reasoning.
4. The tentative behavior model M' is repaired, adapted, and optimized globally.

The following subsection describes the basic elements of this design approach, i. e., Step 2, 3, and 4.

[9] The concepts presented in section have been verified in the hydraulic domain in first place; however, they can be applied to the pneumatic domain in a similar way, suggesting us to use preferably the more generic word "fluidic". Again, a detailed description of the approach can be found in [40].

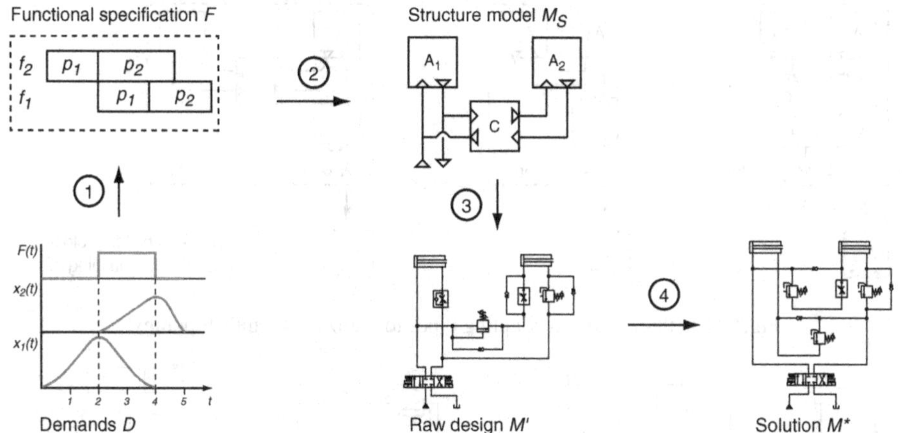

Fig. 8. The functional abstraction paradigm applied to fluidic circuit design.

Remarks. A human designer is capable of working at the component level, *implicitly* creating and combining fluidic axes towards an entire system. His ability to automatically derive function from structure—and vice versa: structure *for* function—allows him to construct a fluidic system without the idea of high-level building blocks.

3.2 Elements of the Design Procedure

Design by functional abstraction rigorously simplifies the underlying domain theory. Here, the tacit assumptions are as follows: (a) Each set of demands, D, can be translated into a set of fluidic functions, F, (b) each function $f \in F$ can be mapped one to one onto a fluidic axis A that operationalizes f, (c) D can be realized by coupling the respective axes for the functions in F, whereas the necessary coupling information can be derived from D.

While the first point goes in accordance with reality, the Points (b) and (c) imply that a function f is neither realized by a combination of several axes nor by constructional side effects. This assumption establishes a significant simplification.

Topology Generation. If the synthesis of fluidic systems is performed at the level of function, the size of the synthesis space is drastically reduced. To be specific, we allow only structure models that can be realized by a recursive application of the three coupling rules shown in Figure 9. The search within this synthesis space is operationalized by means of a design graph grammar [40], which generates reasonable topologies with respect to the functional specification F. The result of this step is a structure model M_S; M_S defines a graph whose nodes correspond to fluidic functions and coupling types.

Case-Based Design of Axes. A structure model M_S is completed towards a behavior model by individually mapping its nodes onto appropriate subcircuits that represent fluidic axes or coupling networks. Figure 10 shows five subcircuits each of which representing a certain fluidic axis. It turned out that the mapping of a fluidic function f onto

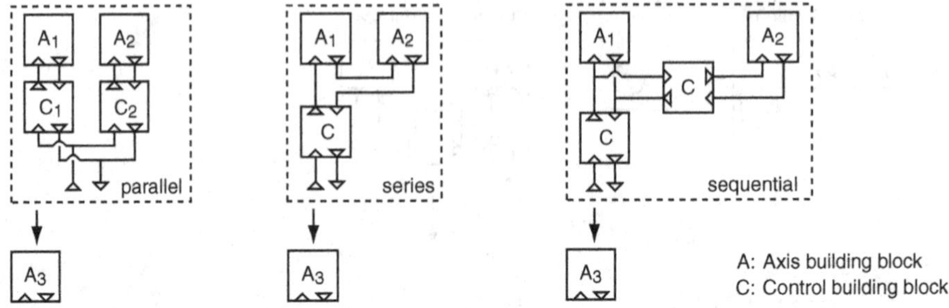

Fig. 9. The three allowed coupling types to realize a circuit's topology.

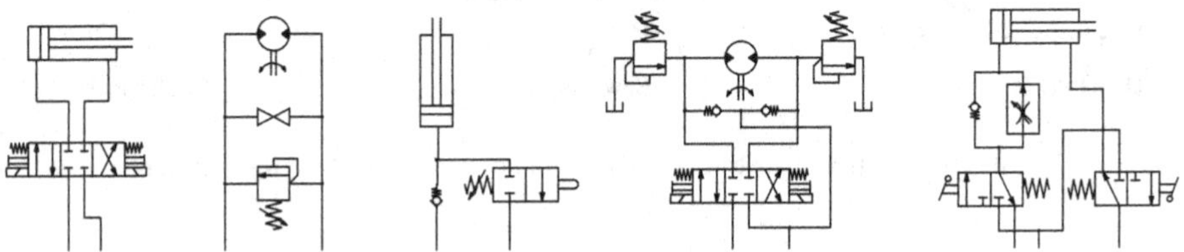

Fig. 10. Five fluidic (hydraulic) axes for different functions and of different complexity.

an axis can be accomplished ideally with case-based reasoning: Domain expert were able to compile a case base $\mathcal{C}$ of about seventy axes that can be used to cover a wide spectrum of fluidic functions.

Speaking technically, an axis is characterized by the unit tasks it can carry out, such as "hold pressure", "fast drive", "hold position", etc. In order to valuate the similarity of fluidic axes and functions, a measure φ was constructed that compares two sequences of unit tasks respecting their types, order, distances, forces, and precision. Moreover, since the axes that are retrieved from $\mathcal{C}$ must be adapted to fulfill a desired f, we also developed a case adaptation scheme that operationalizes engineering know-how in the form of scaling rules. The result of this step, i. e., the composition of adapted fluidic axes according to M_S, yields a preliminary design solution, the raw design M'.

A Design Language for Repair. There is a good chance that a raw design M' has the potential to fulfill D, say, that a sequence of repair steps can be found to transform M' into a behavior model M^*. An example for such a repair measure is the following piece of design knowledge:

"An insufficient damping can be improved by installing a by-pass throttle."

The measure encodes a lot of implicit engineering know-how, among others: (a) A by-pass throttle is connected in parallel, (b) the component to which it is connected is a cylinder, (c) a by-pass throttle is a valve. What is more, the above repair measure can be applied to different contexts in a variety of circuits. To operationalize such kind of knowledge, we developed a prototypic scripting language for fluidic circuit design.[10] In this place we will not delve into language details but refer to [33].

[10] The research was part of the OFT-project, which was supported by DFG grants KL 529/7-1,2,3 and SCHW 120/56-1,2,3.

3.3 Experimental Analysis: Design Automation with ArtDeco-cbd

The outlined concepts have been embedded within the design assistant ArtDeco-cbd, which is linked to a drawing and simulation environment for fluidic systems [19]. The design assistant enables a user to formulate his design requirements as a set of fluidic functions F. For an $f \in F$ a sequence of unit tasks can be defined, where several characteristic parameters such as duration, precision, or maximum values can be stated for each unit task.

Clearly, the crucial question is "How good are the designs of ArtDeco-cbd?" A direct evaluation of the generated models is restricted: an absolute measure that captures the design quality does not exist, and the number of properties that characterizes a design is large and hardly to quantify. On the other hand, the quality of a generated design can be rated *indirectly*, by measuring its "distance" to a design solution created by a human expert. In this connection, the term distance stands for the real modification effort that is necessary to transform the computer solution into the human solution. The experimental results presented in Table 2 report on such a competition.-[11]

Table 2. Runtime and quality results of automatically generated designs. The column "O.K." shows the portion of designs whose simulation fulfills D; only solutions with high similarity values ($\varphi > 0.9$) were considered. The column "Quality" shows the expert evaluation: (+), (o), and (–) indicate a small, an acceptable, and a large modification effort to transform the machine solution into a solution accepted by the human expert.

Number of axes	Time for retrieval	Time for reuse	O.K. (at $\varphi > 0.9$)	Quality		
1	$\ll$ 1s	0.10s	80%	60% (+)	35% (o)	5% (–)
2	$\ll$ 1s	0.63s	75%	50% (+)	45% (o)	5% (–)
3	$\ll$ 1s	0.91s	70%	40% (+)	50% (o)	10% (–)
4	$\ll$ 1s	1.43s	60%	20% (+)	65% (o)	15% (–)
5	$\ll$ 1s	2.00s	20%	5% (+)	80% (o)	15% (–)

Conclusion

The success of a diagnosis or design approach depends on the underlying model space—say, its size, and the way it is explored. A tractable model space is in first place the result of adequate models, which in turn are the result of a skillful selection, combination, and customization of existing construction principles. Engineers don't search because they use adequate models. The challenge is to operationalize such models on a computer.

Especially when expert knowledge is not at hand, the combination of deep models, simulation, and search is inviting because it promises high fidelity. On the other hand, applying a search-plus-simulation paradigm entails the risk to fail completely because of various reasons. This is not inevitable: The principles of model compilation and functional abstraction exemplify how search and simulation can be combined to realize problem solving strategies for complex diagnosis and design problems.

[11] Test environment was a Pentium III system at 450 MHz with 128 MB main memory.

References

1. R. Agrawal, T. Imielinski, and A. Swami. Mining Association Rules between Sets of Items in Large Databases. In Peter Buneman and Sushil Jajodia, editors, *Proceedings of the 1993 ACM SIGMOD International Conference on Management of Data*, Washington D. C., May 1993. ACM Press.
2. Erik K. Antonsson and Jonathan Cagan. *Formal Engineering Design Synthesis*. Cambridge University Press, 2001. ISBN 0-521-79247-9.
3. Paul I. Barton. Modeling, Simulation, and Sensitivity Analysis of Hybrid Systems. In Peter Buneman and Sushil Jajodia, editors, *Proceedings of the IEEE International Symposium on Computer Aided Control System Design*, pages 117–122, Anchorage, Alaska, September 2000. ACM Press.
4. David C. Brown and B. Chandrasekaran. *Design Problem Solving*. Morgan Kaufmann Publishers, 1989.
5. B. Buchanan and E. Shortliffe. *Rule-Based Expert Systems. The MYCIN Experiments of the Stanford Heuristic Programming Project*. Addison-Wesley, Massachusetts, 1984.
6. B. Chandrasekaran and Sanjay Mittal. Deep Versus Compiled Knowledge Approaches to Diagnostic Problem-Solving. In David Waltz, editor, *Proceedings of the National Conference on Artificial Intelligence*, pages 349–354, Pittsburgh, PA, August 1982. AAAI Press. ISBN 0-86576-043-8.
7. William J. Clancey. Heuristic Classification. *Artificial Intelligence*, 27:289–350, 1985.
8. Jean-Marc David, Jean-Paul Krivine, and Reid Simmons, editors. *Second Generation Expert Systems*. Springer, 1992. ISBN 0-387-56192-7.
9. Randall Davis. Expert Systems: Where Are We? And Where Do We Go from Here? *AI Magazine*, 3(2):3–22, 1982.
10. Johan de Kleer. AI Approaches to Troubleshooting. In *Proceedings of the International Conference on Artificial Intelligence in Maintenance*, pages 78–89. Noyes Publications, 1985.
11. Johan de Kleer and Brian C. Williams. Diagnosing Multiple Faults. In M. L. Ginsberg, editor, *Readings in Nonmonotonic Reasoning*, pages 372–388. Morgan Kaufman, 1987.
12. Johan de Kleer and Brian C. Williams. Diagnosis with Behavioral Models. In *Proceedings of the Eleventh International Joint Conference on Artificial Intelligence (IJCAI 89)*, pages 1324–1330, Detroit, Michigan, 1989.
13. Gerhard Fischer, Kumiyo Nakakoji, Jonathan Ostwald, and Gerry Stahl. Embedding Critics in Design Environments. *The Knowledge Engineering Review*, 8(4):285–307, December 1993.
14. Kenneth D. Forbus and Johan de Kleer. *Building Problem Solvers*. MIT Press, Cambridge, MA, 1993. ISBN 0-262-06157-0.
15. Paul Frank. Fault diagnosis: A survey and some new results. *Automatica: IFAC Journal*, 26 (3):459–474, 1990.
16. John S. Gero. Design Prototypes: A Knowledge Representation Scheme for Design. *AI Magazine*, 11:26–36, 1990.
17. W. Hamscher, L. Console, and Johan de Kleer, editors. *Readings in Model-Based Diagnosis*. Morgan Kaufmann, San Mateo, 1992.
18. Sture Hägglund. Introducing Expert Critiquing Systems. *The Knowledge Engineering Review*, 8(4):281–284, December 1993.
19. Marcus Hoffmann. *Zur Automatisierung des Designprozesses fluidischer Systeme*. Dissertation, University of Paderborn, Department of Mathematics and Computer Science, 1999.
20. Uwe Husemeyer. *Heuristische Diagnose mit Assoziationsregeln*. Dissertation (to appear), University of Paderborn, Department of Mathematics and Computer Science, 2001.

21. Werner Karbach and Marc Linster. *Wissensakquisition für Expertensysteme.* Carl Hanser Verlag, 1990. ISBN 3-446-15979-7.
22. David B. Leake. Case-Based Reasoning: Issues, Methods, and Technology, 1995.
23. Eric-Jan Manders, Gautam Biswas, Pieter J. Mosterman, Lee A. Barford, and Robert Joel Barnett. Signal Interpretation for Monitoring and Diagnosis, A Cooling System Testbed. In *IEEE Transactions on Instrumentation and Measurement*, volume 49:3, pages 503–508, 2000.
24. John McDermott. R1: A Rule-based Configurer of Computer Systems. *Artificial Intelligence*, 19:39–88, 1982.
25. Marvin Minsky. Models, Minds, Machines. In *Proceedings of the IFIP Congress*, pages 45–49, 1965.
26. Sriram Narasimhan, Pieter J. Mosterman, and Gautam Biswas. A Systematic Analysis of Measurement Selection Algorithms for Fault Isolation in Dynamic Systems. In *Proc. of the International Workshop on Diagnosis Principles*, pages 94–101, Cape Cod, MA, 1998.
27. Christian J. J. Paredis, A. Diaz-Calderon, Rajarishi Sinha, and Pradeep K. Khosla. Composable Models for Simulation-Based Design. In *Engineering with Computers*, volume 17:2, pages 112–128. Springer, July 2001.
28. Frank Puppe. *Systematic Introduction to Expert Systems, Knowledge Representations and Problem-Solving Methods.* Springer, 1993.
29. Raymond Reiter. A Theory of Diagnosis from First Principles. *Artificial Intelligence*, 32(1): 57–95, April 1987.
30. Michael M. Richter. The Knowledge Contained in Similarity Measures, October 1995. Some remarks on the invited talk given at ICCBR'95 in Sesimbra, Portugal.
31. Michel M. Richter. Introduction to CBR. In Mario Lenz, Brigitte Bartsch-Spörl, Hans-Dieter Burkhard, and Stefan Weß, editors, *Case-Based Reasoning Technology. From Foundations to Applications*, Lecture Notes in Artificial Intelligence 1400, pages 1–15. Berlin: Springer-Verlag, 1998.
32. Michael D. Rychener. *Expert Systems for Engineering Design.* Academic Press, 1988. ISBN 0-12-605110-0.
33. Thomas Schlotmann. Formulierung und Verarbeitung von Ingenieurwissen zur Verbesserung hydraulischer Systeme. Diploma thesis, University of Paderborn, Institute of Computer Science, 1998.
34. L. C. Schmidt and J. Cagan. Configuration Design: An Integrated Approach Using Grammars. *ASME Journal of Mechanical Design*, 120(1):2–9, 1998.
35. Rajarishi Sinha, Christian J. J. Paredis, and Pradeep K. Khosla. Behavioral Model Composition in Simulation-Based Design. In *Proceedings of the 35th Annual Simulation Symposium*, pages 309–315, San Diego, California, April 2002.
36. R. Srikant and R. Agrawal. Mining Quantitative Association Rules in Large Relational Tables. In H. V. Jagadish and I. S. Mumick, editors, *Proceedings of the 1996 ACM SIGMOD International Conference on Management of Data*, pages 1–12, Montreal, Canada, June 1996. ACM Press.
37. Luc Steels. Components of Expertise. *AI Magazine*, 11(2):28–49, 1990.
38. Benno Stein. *Functional Models in Configuration Systems.* Dissertation, University of Paderborn, Institute of Computer Science, 1995.
39. Benno Stein. Generating Heuristics to Control Configuration Processes. *Applied Intelligence, APIN-IEA, Kluwer*, 10(2/3):247–255, March 1999.
40. Benno Stein. *Model Construction in Analysis and Synthesis Tasks.* Habilitation thesis, University of Paderborn, Institute of Computer Science, 2001.

41. Benno Stein. Model Compilation and Diagnosability of Technical Systems. In *Proceedings of the third IASTED International Conference on Artificial Intelligence and Applications (AIA 03)*, Benalmádena, Spain, 2003.
42. Peter Struß and Oskar Dressler. "Physical Negation"—Integrating Fault Models into the General Diagnostic Engine. In *Proceedings of the Fifteenth International Joint Conference on Artificial Intelligence (IJCAI 89)*, volume 2, pages 1318–1323, Detroit, Michigan, USA, 1989.

Randomized Search Heuristics as an Alternative to Exact Optimization

Ingo Wegener*

FB Informatik, LS2, Univ. Dortmund, 44221 Dortmund, Germany
ingo.wegener@uni-dortmund.de

Abstract. There are many alternatives to handle discrete optimization problems in applications. Problem-specific algorithms vs. heuristics, exact optimization vs. approximation vs. heuristic solutions, guaranteed run time vs. expected run time vs. experimental run time analysis. Here, a framework for a theory of randomized search heuristics is presented. After a brief history of discrete optimization, scenarios are discussed where randomized search heuristics are appropriate. Different randomized search heuristics are presented and it is argued why the expected optimization time of heuristics should be analyzed. Afterwards, the tools for such an analysis are described and applied to some well-known discrete optimization problems. Finally, a complexity theory of so-called black-box problems is presented and it is shown how the limits of randomized search heuristics can be proved without assumptions like NP $\neq$ P. This survey article does not contain proofs but hints where to find them.

1 Introduction

For our purposes it makes no sense to look for the ancient roots of algorithmic techniques. The algorithmic solution of large-scale problems was not possible before computers were used to run the algorithm. Some of the early algorithms of this period like Dantzig's simplex algorithm for linear programming from 1947 and Ford and Fulkerson's network flow algorithm from 1956 based on improvements along augmenting paths were quite successful. In the fifties and early sixties of the last century one was satisfied when the algorithm was running efficiently – in most experiments. Nowadays, we know that the simplex algorithm has an exponential worst-case run time and that small randomized pertubations in the input turn all inputs into easy ones with respect to the input distribution based on the pertubation (Spielman and Teng (2001)). The first network flow algorithm was only pseudo-polynomial. Now, we know of many polynomial network flow algorithms. However, the original simplex algorithm still finds

* Supported in part by the Deutsche Forschungsgemeinschaft (DFG) as part of the Collaborative Research Center "Computational Intelligence" (SFB 531) and by the German Israeli Foundation (GIF) in the project "Robustness Aspects of Algorithms".

W. Lenski (Ed.): Logic versus Approximation, LNCS 3075, pp. 138–149, 2004.

many applications and the best network flow algorithms use ideas of Ford and Fulkerson.

Later, algorithm design was accompanied by a run time analysis. Many algorithms have good worst-case run times. Randomized algorithms are often simpler and faster than deterministic ones. The NP-completeness theory proves that there are many important problems which are intractable – with respect to the worst case run time (see Garey and Johnson (1979)). At the same time, approximation algorithms guaranteeing good solutions in short time were presented and analyzed (see Hochbaum (1997)). All the experts believe that NP $\neq$ P ant it makes sense to argue under this hypothesis. In the seventies many people believed that we cannot solve NP-hard optimization problems for large instances. As computers got faster, applications were showing that this belief is wrong. There are problem-specific algorithms solving the "typical cases" of difficult problems sufficiently fast. In most cases, it was impossible to describe the "easy instances" for such algorithms. Nowadays, we have a tool-box of algorithmic techniques to design such problem-specific algorithms. Among them are greedy algorithms, dynamic programming, branch-and-bound, relaxation techniques, and many more. Randomized algorithms are now very common and even derandomization techniques exist. The community working on the design and analysis of algorithms has broadened its scope. The experimental tuning of algorithms named algorithm engineering is part of this discipline. However, the community is still focused on problem-specific algorithms.

Randomized search heuristics (RSH) in their pure form are algorithmic techniques to attack optimization problems with one general strategy. They are heuristics in a strong sense. Typically, they do not guarantee any (non-trivial) time bound and they do not guarantee any quality of the result. Randomized local search (RLS) is a member of this family of algorithms if it does not use problem-specific neighborhoods. Simulated annealing (SA) and all kinds of evolutionary algorithms (EA) (including evolution strategies (ES) and genetic algorithms (GA)) are typical RSHs. The idea was to simulate some successful strategy of engineering (annealing) or nature (evolution). The history of these RSHs dates back to the fifties and sixties of the last century. Despite some early successes they did not find so many applications before the late eighties since computers were not fast enough. Nowadays, people in applications like these heuristics and people in the algorithm community have ignored them for a long time. Indeed, these RSHs were considered as the black sheeps in the algorithm community. There were several reasons for this. RSHs were often claimed to be the best algorithm for a problem which was wrong because of very clever problem-specific algorithms. Arguments supporting in particular EAs were based on biology and not on a run time analysis. So there were many "soft" arguments and the two communities were using different languages. The approach to analyze RSHs like all other algorithms is a bridge between the two communities and both of them have started to use this bridge.

In this overview, we try to describe this approach. First, it is important to discuss the scenarios where we recommend RSHs in their pure form. This is necessary to avoid many common misunderstandings, see Section 2. In Section 3, we present the general form of a black-box algorithm and discuss some types of SA and EA. As already described, we propose the run time analysis of RSHs. In Section 4, we discuss what we expect from such an approach and compare this approach with the other theoretical results on RSHs. In Section 5, we describe the tool-box for the analysis of RSHs and, in Section 6, we present some results in order to show that we can realize our goals – at least in some cases. In the classical scenario, design and analysis of algorithms is accompanied by complexity theory showing the limits of the algorithmic efforts. In Section 7, we argue that an information-restricted scenario, namely the scenario of black-box algorithms, can be used to prove some limits of RSHs. Since the available information is limited, these limits can be proved without complexity theoretical hypotheses like NP $\neq$ P. We finish with some conclusions.

2 Scenarios for the Application of Randomized Search Heuristics

The first fact we have to accept is that no algorithmic tool will be the best for all problems. We should not expect that some RSH will beat quicksort or Dijkstra's shortest path algorithm. Besides these obvious examples there are many problems where the algorithm community has developed very efficient algorithms. The following statement seems to be true in almost all situations. If a problem has a structure which is understood by the algorithm community and if this community has made some effort in designing algorithms for the problem, then an RSH in its pure form will be worse with respect to its run time and/or the quality of its results. Many people working on RSHs do not agree with this statement. This is due to a misunderstanding. The statement argues about RSHs in their pure form. If they have problem-specific components, they can be competitive but then they are already hybrid algorithms combining ideas from RSHs with problem-specific ideas. If such hybrid algorithms are possible, one should consider them as alternative. For hybrid algorithms, it is difficult to estimate whether the general idea of an RSH or a problem-specific module is the essential part.

The question is whether there are scenarios where one should apply RSHs in their pure form. We think of two such scenarios.

In many projects, optimization problems are subproblems and a good algorithm has to be presented in short time. If no expert has considered the problem before and if there are no experts in algorithm design in the project and if there is not enough time and/or money to ask a group of experts, then an RSH can be a good choice. In real applications, the small amount of knowledge about the problem, should be used. RSHs have turned out to be a robust tool which for many problems leads to satisfactory results. In this scenario, we expect that better algorithms exist but are not available. In many applications, there is no knowledge about the problem. As an example, think of the engineering problem to "optimize some machinery". The engineer can choose the values of certain free parameters and each choice "realizes a machine". There exists a function $f\colon M \to \mathbb{R}$ where $f(x)$ is the quality of the machine $x \in M$. Because of the complexity of the application, the function f is not known and $f(x)$ has to be estimated by an experiment or its simulation with a computer. Here, a problem-specific approach is impossible and RSHs are a good choice.

Summarizing, there are scenarios where RSHs are applied in (almost) pure form. Whenever possible, one should apply all available problem-specific knowledge.

3 How Randomized Search Heuristics Work

We consider the maximization of some function $f\colon S \to \mathbb{R}$ where S is a finite (discrete) search space. The main difference between an RSH and a problem-specific algorithm is the following. The RSH does not assume knowledge about f. It may sample some information about f by choosing search points $x \in S$ and evaluating $f(x)$. Therefore, we may think of a black box containing f and the algorithm not knowing f can use the black box in order to get the answer $f(x)$ to the query x. Hence, the general form of a black-box algorithm can be described as follows.

The General Black-Box Algorithm (BBA)

The initial information I_0 is the empty sequence. In the tth step, based on $I_{t-1} = ((x_1, f(x_1)), \dots, (x_{t-1}, f(x_{t-1})))$, the algorithm computes a probability distribution p_t on S, chooses x_t according to p_t, and uses the black box to obtain $f(x_t)$. If a stopping criterion is fulfilled, the search is stopped and a search point x_i with maximal f-value is presented as result.

All known RSHs fit into this framework. Most often, they are even more specialized. The information I_t is considered as (unordered) multiset and not all the information is kept in the storage. There is a bound $s(n)$ on the number of pairs $(x, f(x))$ which can be stored and in order to store the new pair $(x_t, f(x_t))$ one has to throw away another one. This is called an $s(n)$-BBA. RLS and SA work with $s(n) = 1$ and, for EAs, $s(n)$ usually is not very large.

If $s(n) = 1$, we need a search operator which produces a new search point x' depending on the current search point x and its f-value. Then we need a selection procedure which decides whether x' replaces x. RLS and SA restrict the search to a small neighborhood of x. SA allows that worse search points replace better ones. The special variant called Metropolis algorithms accepts x' with a probability of $\exp(-(f(x) - f(x'))/T)$ for some fixed parameter $T \geq 0$, called the temperature, if $f(x') < f(x)$, and it accepts x', if $f(x') \geq f(x)$. SA varies T using a so-called cooling schedule. This schedule depends on t and we have to store the point of time besides the search point. EAs prefer more global search operators. A so-called mutation step on $S = \{0,1\}^n$ flips each bit independently with a probability p_n, its typical calue is $1/n$. Because of this global search operator EAs can decide to accept always the best search points (the so-called plus strategy). The $s(n)$ search points kept in the storage are called population and it is typical to produce more than one search point before selecting those which survive. These phases are called generations. There are many selection procedures that we do not have to discuss in detail. If the population size is larger than 1, we also need a selection procedure to select those search points (called individuals) that are the objects of mutation. Finally, larger populations allow search operators depending on more than one individual. Search operators working on one individual are called mutation and search operators working on at least two (in most cases exactly two) individuals are called crossover or recombination. This leads to many free parameters of an EA and all the parameters can be changed over time (dynamic EAs for a fixed schedule and adaptive EAs otherwise). In the most general form of self-adaptive EAs, the free parameters are added to the search points (leading to a larger search space) and are changed by mutation and recombination. The class of pure EAs is already large and there is the freedom of adding problem-specific modules.

4 Arguments for the Analysis of the Expected Optimization Time

Do we need a theory of RSHs? They are heuristics – so let us try them! There is a common belief that a theory would improve the understanding of RSHs and, therefore, the choice of the parameters and the application of RSHs. Theory is contained in all the monographs on RSHs. The question is what type of theory is the right choice. The following discussion presents the different types of theoretical approaches (with an emphasis on EA theory) and some personal criticism in order to motivate our approach.

In general, we investigate a class of functions or, equivalently, a problem consisting of a class of instances. Many RSHs are quite complex but even simple variants like the Metropolis algorithm or a mutation-based EA with population size 1 have a complex behavior for most problems. People in classical algorithm theory design an algorithm with the aim that the algorithm is efficient and the aim that they are able to prove this. Here, the algorithm realizes a general search strategy which does not support its analysis. Thus, the analysis of RSHs has to overcome different obstacles than classical algorithm analysis.

This has led many researchers to simplify the situation by analyzing a so-called model of the algorithm. Afterwards, experiments "verify" the quality of the model. This procedure is common in physics. There, a model of "nature" is necessary since "reality" cannot be described accurately. An algorithm and, therefore, an RSH is nothing from nature but an abstract device made by humans. It is specified exactly and, ignoring computer failures, we can analyze the "true algorithm" and a model is not necessary. An analysis of the algorithm can give precise results (at least in principle) and this makes a verification superfluous. Moreover, the dimension of our "world", namely the search space, is not limited.

Experiments can tell us only something about those problem dimensions n that we can handle now. A theoretical analysis can give theorems for all n.

In order to handle large populations, people have studied the model of infinite populations (described by probability vectors) which are easier to handle. There are only few papers investigating the error of this model with respect to a restricted population size and time (Rabani, Rabinovich, and Sinclair (1998) and Ollivier (2003)).

An RSH is a stochastic process and one can study the dynamics of this process. Such results are often on a high technical level but the results are hard to interpret if we are interested in the process as an optimization algorithm.

There are also many papers analyzing quite precisely the result of one step of the algorithm. How much do we improve the best known f-value? How close (in distance in the search space) do we approach an optimal search point? The complete knowledge of the one-step behavior is (in principle) enough to derive the complete knowledge of the global behavior of the process. However, this derivation is the hard step. As known from classical algorithm analysis, we should be happy to understand the asymptotic behavior of algorithms and, for this, it is not necessary to have a precise understanding of the one-step behavior. Well-known concepts as schema theory or building block hypothesis have led people to wrong conjectures about algorithms.

Finally, there are many papers proving the convergence of RSHs to the optimum. Without an estimate of the speed of convergence, these results are of limited interest – at least in finite search spaces. Convergence is guaranteed easily if we store the best search point ever seen. Then an enumeration of the search space suffices as well as a mutation operator giving a positive probability to reach every search point.

Our approach is to analyze an RSH as any other randomized algorithm. The only difference is that an RSH cannot know whether it has found an optimal search point. In most applications, the stopping criterion is not a big problem. Therefore, we investigate an RSH as an infinite stochastic process (without stopping criterion) and analyze random variables of interest.

A typical example is the random variable describing the first point of time where an optimal search point is passed to the black box. Its expected value is called the expected run time or optimization time of the algorithm. In order to analyze restart techniques, it is also interesting to analyze the success probability, namely the probability of the event that the run time is bounded by some given time bound. We have to be careful since we only count the number of search points and, therefore, the number of f-evaluations, and not the effort to compute the search points. For most RSHs, this is a good measure. If one applies time-consuming procedures to compute search points, one has to analyze also these resources.

5 Tools for the Analysis of Randomized Search Heuristics

As in classical algorithm analysis, one has to develop a good intuition how the given RSH works for the considered problem. Then one can describe "typical runs" of the RSH. They are approaching the optimum by realizing sequentially certain subgoals. One tries to estimate the run time for reaching subgoal i assuming that subgoal $i-1$ has been realized. This can lead to a time bound $t_i(n)$ for Phase i. The aim is to estimate the error probability $\varepsilon_i(n)$ that Phase i is not finished within $t_i(n)$ steps. The sum of all $\varepsilon_i(n)$ is an upper bound on the probability that the run is not typical, e. g., that it takes longer than the sum of all $t_i(n)$. Obviously, the last subgoal has to be the goal to find an optimal search point. Often it is possible to obtain exponentially small $\varepsilon_i(n)$ implying that even polynomially many subgoals lead to an exponentially small error probability. (See Jansen and Wegener (2001) for a typical application of this method.)

A related proof technique is to define an f-based partition $(A_1, \ldots, A_m)$ of the search space, i. e. $x \in A_i$, $y \in A_j$, and $j > i$ imply $f(x) < f(y)$ in the case of maximization and A_m contains exactly the optimal search points. Then one has to estimate the expected time $t_i(n)$ of finding a search point in some A_j, $j > i$, given that a search point in A_i is known. Here we have to assume that we never forget the best of the search points ever seen.

As already mentioned, the stochastic process P describing an RSH can be quite complex. A simple and powerful idea (which typically is difficult to realize) is to construct a stochastic process P' which is provably slower than P but easier to analyze than P. In order to obtain good bounds, P' should not be "much slower" than P. The notion slower can be defined as follows. A random variable X is called not larger than X' if $\text{Prob}(X \le b) \ge \text{Prob}(X' \le b)$ for all b. The process P is not faster than P' if the random variable T describing a good event for P is not larger than the corresponding random variable T' for P'. (See Wegener and Witt (2003) for a typical application of this method.)

Finally, it is not always the right choice to measure the progress of the RSH with respect to the given function f (often called fitness function) but with respect to some pseudo-fitness function g known in classical algorithm design as potential function. Note that the algorithm still works on f and g is only used in the analysis of the algorithm. (See Droste, Jansen, and Wegener (2002) for a typical application of this method.)

During all the calculations, one needs technical tools from probability theory, among them all the famous tail inequalities (due to Markoff, Tschebyscheff, Chernoff, or Hoeffding). Other tools are the coupon collector's theorem or results on the gambler's ruin problem.

Finally, we mention a special technique called delay sequence arguments. This technique has been introduced by Ranade (1991) for routing problems and has been applied only once for RSHs (Dietzfelbinger, Naudts, van Hoyweghen, and Wegener (2002)). The idea is the following. If we expect that a stochastic process is finished in time $t(n)$ with large probability, we look for an event which has to happen if the stochastic process is delayed, i. e., if it runs for more than

$t(n)$ steps. Then we estimate the probability of the chosen event. Despite of the simplicity of the general idea, this method has been proven to be quite powerful.

Most RSHs apply in each step the same techniques of search and selection. This implies that the same experiment is applied quite often. The experiments are influenced by the available information I_t but in many aspects these experiments are independent or almost independent. Because of the large number of experiments during a search or a subphase of a search, many interesting random variables are highly concentrated around their expected values. This simplifies the analysis and can lead to sharp estimates.

6 The Analysis of Randomized Search Heuristics on Selected Problems

The analysis of the expected optimization time of RSHs is far behind the analysis of classical algorithms. One reason is that the number of researchers in this area is still much smaller and that they have started to work on this subject much later. It will take several years until the research on the analysis of RSHs can reach the status of classical algorithm analysis. Nevertheless, it is not possible to give here an overview of all the results obtained so far. We discuss some results to motivate the reader to take a closer look to the original papers and this section has a focus on results obtained in our research group.

In order to build a bridge between the RSH community gathering in conferences like GECCO, PPSN, CEC, and FOGA and the classical algorithm community gathering in conferences like STOC, FOCS, ICALP, STACS, SODA, and ESA, we have tried to work on topics of the following kind:

- solving open problems on RSHs discussed in the RSH community,
- analyzing RSHs on functions which are considered as typical in the RSH community,
- analyzing RSHs on classes of functions described by structural properties, and
- analyzing RSHs on well-known problems of combinatorial optimization.

Concerning EAs, the American school (Holland (1975), Goldberg (1989), Fogel (1995)) proposed GAs where crossover is the most important search operator and the European school (Rechenberg (1994), Schwefel (1995)) proposed ESs merely based on mutation. Concerning combinatorial optimization, several results have shown that mutation is essential. What about crossover? Even after 30 years of debates there was no example known where all mutation-based EAs have a super-polynomial optimization time while a generic GA works in expected polynomial time. Jansen and Wegener (2002), the conference version appeared 1999, were the first to present such a result. The considered function is quite simple but the analysis is complicated. Crossover can be useful only between two quite different search points. Selection prefers good search points which can decrease the diversity of the population. The GA which has been analyzed does not use special modules to preserve the diversity and one has to prove that the population nevertheless is diverse enough. This paper proves a trade-off of super-polynomial versus polynomial. This has been improved in a later paper (Jansen and Wegener (2001)) to exponential versus polynomial. This is due to the definition of an artificial function which supports the analysis of the GA.

In this survey article, it does not make sense to introduce functions that have been discussed intensively in the RSH community.

Each pseudo-boolean function $f\colon \{0,1\}^n \to \mathbb{R}$ can be described uniquely as a polynomial

$$f(x) = \sum_{A \subseteq \{1,\ldots,n\}} w_A \prod_{i \in A} x_i$$

Its degree is the maximal $|A|$ where $w_A \neq 0$. We can consider functions with bounded degree. Degree-1 functions, also called linear functions, were investigated in many papers. In particular, the expected run time of the (1+1) EA (population size 1, mutations flipping the bits independently with probability $1/n$, replacing x by x' if $f(x') \geq f(x)$ in the case of maximization) was of interest. Droste, Jansen, and Wegener (2002) have proved that the expected run time is always $O(n \log n)$ and $\Omega(n \log n)$ if $\Omega(n^\varepsilon), \varepsilon > 0$, weights are non-zero. The proof applied the techniques of potential functions and the investigation of slower stochastic processes.

The maximization of degree-2 or quadratic polynomials is NP-hard. Wegener and Witt (2002) have investigated RSHs on certain quadratic functions. They have proved that restart techniques can decrease the expected optimization time from exponential to polynomial and they have presented a quadratic polynomial where all mutation-based EAs are slow. Moreover, they have investigated the case of monotone quadratic polynomials. A polynomial is called monotone if it can be represented with non-negative weights after replacing some x_i by $1 - x_i$. This allows that the polynomial is monotone increasing with respect to some variables and monotone decreasing with respect to the other variables. Wegener and Witt (2003) have generalized this approach to degree-d polynomials. For RLS and the (1+1) EA with a small probability of flipping bits they have proved an upper bound of $O(2^d \cdot (n/d) \cdot \log(n/d))$ which is optimal since there is a corresponding lower bound for the sum of all $x_{id+1} \cdots x_{(i+1)d}$, $0 \leq i \leq \lfloor n/d \rfloor - 1$. The proof of the upper bound is mainly based on the investigation of slower stochastic processes.

Finally, there is some recent progress in the analysis of EAs on combinatorial optimization problems. Many earlier results consider RLS and some results consider SA. Scharnow, Tinnefeld, and Wegener (2002) have shown that the complexity of the problem depends on the choice of the corresponding fitness function. The sorting problem is the best investigated computer science problem. It can be considered as the problem of minimizing the unsortedness. There are many measures of unsortedness known from the design of adaptive sorting algorithms. E. g., we count the number of inversions, i. e., of pairs of objects in incorrect order or we count the number of runs or, equivalently, the number of adjacent pairs of objects in incorrect order. A variant of the (1+1) EA working on the search space of all permutations can sort in expected time $O(n^2 \log n)$ using the number of inversions (and several other measures of unsortedness) but it needs exponential time using the number of runs. The same effect has been shown in the same paper for the single-source-shortest-paths problem. The search points are directed trees rooted at the source and, therefore, containing paths to all other vertices. If the fitness is described by the sum of the lengths of the paths from the source to all other vertices, each black-box search heuristic needs exponential time while time $O(n^3)$ is enough on the average if the fitness is described by the vector containing all the path lengths.

The (1+1) EA is surprisingly efficient for the minimum spanning tree problem (Neumann and Wegener (2003)). For graphs on n vertices and m edges and edge weights bounded by $w_{\max}$, the expected run time is bounded by $O(m^2(\log n + \log w_{\max}))$.

The maximum matching problem is more difficult for RSHs. There are example graphs where SA (Sasaki and Hajek (1988)) and the (1+1) EA (Giel and Wegener (2003)) need expected exponential time. In the latter paper it is shown that the (1+1) EA is efficient on certain simple graphs. More important is the following result. The (1+1) EA is a PRAS (polynomial-time randomized approximation scheme), i. e., an approximation ratio of $1+\varepsilon$ can be obtained in expected polynomial time where the degree of the polynomial depends on $1/\varepsilon$. Due to the examples mentioned above this is the best possible result. The true aim of search heuristics is to come close to the optimum, exact optimization is not necessary.

7 Black-Box Complexity – The Complexity Theoretical Background

The general BBA works in an information-restricted scenario. This allows the proof of lower bounds which hold for any type of BBA. This theory based on Yao's minimax principle (Yao (1977)) has been developed and applied by Droste, Jansen, and Wegener (2003). A randomized BBA is nothing but a probability distribution on the set of deterministic BBAs. This is obvious although it can be difficult to describe this probability distribution explicitly if a randomized BBA is given. For a deterministic BBA storing the whole history it does not make sense to repeat a query asked earlier. For a finite search space, the number of such deterministic BBAs is finite. In many cases, also the set of problem instances of dimension n is finite.

In such a situation, Yao's minimax principle states that we obtain a lower bound on the expected optimization time of each randomized BBA by choosing a probability distribution on the set of problem instances and proving a lower bound on the average optimization time of each *deterministic* BBA where the average is taken with respect to the random problem instance. The key idea is to consider algorithm design as a zero-sum game between the algorithm designer and the adversary choosing the problem instance. The algorithm designer has to pay 1 Euro for each query. Then Yao's minimax principle is a corollary to the minimax theorem for two-person zero-sum games.

Applying this theory leads to several interesting results. It is not too hard to prove that randomized BBAs need $\Omega(2^d + n/\log n)$ queries on the class of monotone d-degree polynomials proving that RLS and the considered variant of (1+1) EA are close to optimal for these functions. Also, the exponential lower bound for the single-source-shortest-paths problem follows from this theory.

Many people have stated that EAs are particularly efficient on unimodal functions $f\colon \{0,1\}^n \to \mathbb{R}$. Such functions have a unique global optimum and no further local optimum with respect to the neighborhood of Hamming distance 1. We investigate unimodal functions $f\colon \{0,1\}^n \to \{0,\dots,b\}$ with a bounded image set. The upper bound for EAs is the trivial bound $O(nb)$ which also can be realized by deterministic local search. This bound indeed is not far from optimal. In a different framework, Aldous (1983) has investigated one-to-one unimodal functions and has proved that their black-box complexity is $2^{n/2}$ up to polynomial factors. In applications, the case of limited b is more interesting. Droste, Jansen, and Wegener (2003) have proven a lower bound of $\Omega(b/\log^2 b)$ if $2n \le b = 2^{o(n)}$.

Sometimes, upper bounds on the black-box complexity are surprisingly small. This happens if the algorithm uses much time to evaluate the information contained in I_t. This can lead to polynomially many queries for NP-equivalent problems. These algorithms use exponential time to evaluate the information. If we restrict this time to polynomial, lower bounds are based on hypotheses like NP$\neq$P. It is an open problem to prove in such cases exponential lower bounds for $s(n)$-BBAs and small $s(n)$.

Conclusions

Applications in many areas have proved that RSHs are useful in many situations. A theory on RSHs in the style of the theory on problem-specific algorithms will improve the understanding of RSHs and will give hints how to choose the free parameters depending on the problem structure. Moreover, RSHs will be included in lectures on efficient algorithms only if such a theory is available. It has been shown that such a theory is possible and that first results of such a theory have been obtained.

References

1. Aldous, D. (1983). Minimization algorithms and random walk on the d-cube. The Annals of Probability 11, 403–413.
2. Dietzfelbinger, M., Naudts, B., van Hoyweghen, C., and Wegener, I. (2002). The analysis of a recombinative hill-climber on H-IFF. Accepted for publication in IEEE–Trans. on Evolutionary Computation.
3. Droste, S., Jansen, T., and Wegener, I. (2002). On the analysis of the (1+1) evolutionary algorithm. Theoretical Computer Science 276, 51–81.

4. Droste, S., Jansen, T., and Wegener, I. (2003). Upper and lower bounds for randomized search heuristics in black-box optimization. Accepted for publication in Theory of Computing Systems.
5. Fogel, D. B. (1995). *Evolutionary Computation: Toward a New Philosophy of Machine Intelligence.* IEEE Press, Piscataway, NJ.
6. Garey, M. R. and Johnson, D. B. (1979). *Computers and Intractability. A Guide to the Theory of NP-Completeness.* W. H. Freeman.
7. Giel, O. and Wegener, I., (2003). Evolutionary algorithms and the maximum matching problem. Proc. of 20th Symp. on Theoretical Aspects of Computer Science (STACS), LNCS 2607, 415–426.
8. Goldberg, D. E. (1989). *Genetic Algorithms in Search, Optimization, and Machine Learning.* Addison-Wesley, Reading, MA.
9. Hochbaum, D. (ed.) (1997). *Approximation Algorithms for NP-Hard Problems.* PWS Publishing Company, Boston.
10. Holland, J. H. (1975). *Adaptation in Natural and Artificial Systems.* Univ. of Michigan, MI.
11. Jansen, T. and Wegener, I. (2001). Real royal road functions – where crossover is provably essential. Proc. of the Genetic and Evolutionary Computation Conference (GECCO '2001). Morgan Kaufmann, San Mateo, CA, 375–382.
12. Jansen, T. and Wegener, I. (2002). The analysis of evolutionary algorithms – a proof that crossover really can help. Algorithmica 34, 47–66.
13. Neumann, F. and Wegener, I. (2003). Randomized local search, evolutionary algorithms, and the minimum spanning tree problem. Submitted for publication.
14. Ollivier, Y. (2003). Rate of convergence of crossover operators. Random Structures and Algorithms 23, 58–72.
15. Rabani, Y. Rabinovich, Y., and Sinclair, A. (1998). A computational view of population genetics. Random Structures and Algorithms 12, 314–330.
16. Ranade, A. G. (1991). How to emulate shared memory. Journal of Computer and System Sciences 42, 307–326.
17. Rechenberg, I. (1994). *Evolutionsstrategie '94.* Frommann-Holzboog, Stuttgart.
18. Sasaki, G. H. and Hajek, B. (1988). The time complexity of maximum matching by simulated annealing. Journal of the ACM 35, 387–403.
19. Scharnow, J., Tinnefeld, K., and Wegener, I. (2002). Fitness landscapes based on sorting and shortest paths problems. Proc. of 7th Conf. on Parallel Problem Solving from Nature (PPSN-VII), LNCS 2439, 54–63.
20. Schwefel, H.-P. (1995). *Evolution and Optimum Seeking.* Wiley, New York.
21. Spielman, D. A. and Teng, S.-H. (2001). Smoothed analysis of algorithms: why the simplex algorithm usually takes polynomial time. Proc. of 33rd ACM Symp. on Theory of Computing (STOC), 296–305.
22. Wegener, I. and Witt, C. (2002). On the analysis of a simple evolutionary algorithm on quadratic pseudo-boolean functions. Accepted for publication in Journal of Discrete Algorithms.
23. Wegener, I. and Witt, C. (2003). On the optimization of monotone polynomials by simple randomized search heuristics. Accepted for publication in Combinatorics, Probability and Computing.
24. Yao, A. C. (1977). Probabilistic computations: Towards a unified measure of complexity. Proc. of 17th IEEE Symp. on Foundations of Computer Science (FOCS), 222–227.

Approximation of Utility Functions by Learning Similarity Measures

Armin Stahl

University of Kaiserslautern, Computer Science Department
Artificial Intelligence - Knowledge-Based Systems Group
67653 Kaiserslautern, Germany
stahl@informatik.uni-kl.de

Abstract. Expert systems are often considered to be logical systems producing outputs that can only be correct or incorrect. However, in many application domains results cannot simply be distinguished in this restrictive form. Instead to classify a result as correct or incorrect, here results might be more or less *useful* for solving a given problem or for satisfying given user demands, respectively. In such a situation, an expert system should be able to estimate the *utility* of possible outputs a-priori in order to produce reasonable results. In Case-Based Reasoning this is done by using similarity measures which can be seen as an approximation of the domain specific, but a-priori unknown *utility function.* In this article we present an approach how this approximation of utility functions can be facilitated by employing machine learning techniques.

1 Introduction

When developing knowledge-based systems one must be aware of the fact that the output of such a system often cannot simply be judged as correct or incorrect. Instead, the output may be more or less *useful* for solving a given problem or for satisfying the users' demands. Of course, one is interested in maximizing the *utility* of the output even if the underlying *utility function* is (partially) unknown. Therefore, one must at least be able to estimate the a-priori unknown utility of possible outputs, for example, by employing heuristics.

Typical examples for knowledge-based systems where we have to deal with unknown utility functions are *Information Retrieval (IR)* [12] and *Case-Based Reasoning (CBR)* [1,9] systems. In IR systems one is interested in finding textual documents that contain information that is useful for satisfying the information needs of the users. Since CBR systems are used for various application tasks, here, we have to distinguish different situations. On the one hand, for traditional application fields like classification, the external output of a CBR system — here the predicted class membership of some entity — usually can only be correct or incorrect, of course. However, due to the problem solving paradigm of CBR, internally a case-based classification system relies on an appropriate estimation

W. Lenski (Ed.): Logic versus Approximation, LNCS 3075, pp. 150–172, 2004.

of the utility of *cases*[1] which are used to infer the class prediction. On the other hand, in more recently addressed application fields, also the external outputs of CBR systems underlie utility functions. A typical example are product recommendation systems used in e-Commerce [4,6,19]. Here, CBR systems are used to select products (or services) — represented as cases — that fulfill the demands and wishes of customers as good as possible. Hence, a recommended product does not represent a correct or incorrect solution, but a more or less suitable alternative.

In CBR the utility of cases is estimated according the following heuristics: *"Similar problems have similar solutions"*. Here, the assumption is that a case consists of a problem description and the description of a corresponding, already known solution. When being confronted with a new problem for which a solution is required, the description of this problem — also called *query* — has to be compared with the problem descriptions contained in available cases. If two problems are *similar* enough, the probability that also similar solutions can be applied to both problems should be high. This means, the similarity between a given problem situation and the problem described in a case can be seen as a heuristics to estimate the utility of the corresponding known solution for solving the current problem (see Figure 1). If the found solution cannot directly be reused to solve the given problem, it has to be *adapted* so that it fits the changed requirements.

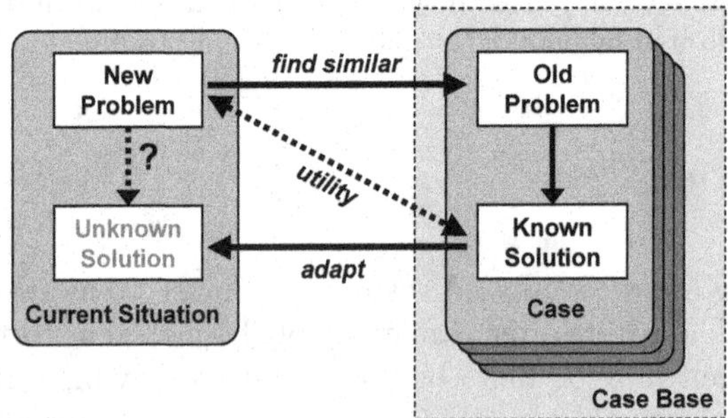

Fig. 1. Approximating Utility through Similarity in CBR

In order to be able to calculate the similarity of two problem descriptions, so-called *similarity measures* are employed. Basically, a similarity measure can be characterized as an approximation of an a-priori unknown utility function [3]. The quality of this approximation strongly depends on the amount of domain specific knowledge one is able to encode into the similarity measure. However, a manual definition of *knowledge-intensive similarity measures* leads to the well known *knowledge acquisition bottleneck* when developing knowledge-based sys-

[1] In CBR cases typically represent experiences about already solved problems or other information that can be reused to solve a given problem.

tems. In this article we show that machine learning approaches can be applied to learn similarity measures from a special kind of (user) feedback, leading to better approximations of the underlying utility functions. A more detailed description of the presented approach is given by [17].

In Section 2 we first introduce the foundations of similarity measures in general and we show how they are represented in practice. In Section 3 we present our approach to learning similarity measures which is based on a special kind of feedback and corresponding learning algorithms. To demonstrate the capabilities of our approach, in Section 5 we describe some evaluation experiments and discuss the corresponding results. Finally, we close with a short summary and conclusion.

2 Similarity Measures

Basically, the task of a similarity measure is to compare two cases[2], or at least particular parts of two cases, and to compute a numeric value which represents the degree of similarity. In traditional CBR systems the cases' parts to be compared are usually descriptions of problem situations, however, in more recent application domains a clear distinction between problems and solutions is often not given. Hence, in the following we assume that a similarity measure compares two *case characterizations* describing the aspects that are relevant to decide whether a case might be useful for a given query or not:

Definition 1 (Similarity Measure, General Definition). *A* ***similarity measure*** *is a function* $Sim : \mathbb{D} \times \mathbb{D} \longrightarrow [0,1]$*, where* $\mathbb{D}$ *denotes the space of case characterizations.*

Depending on the application tasks, case characterizations might be represented by using very different formalisms. Of course, the used formalism strongly influences the manner how corresponding similarity measures have to be represented. Therefore, we first introduce the kind of case representation that we presume in the following.

2.1 Attribute-Value Based Case Representation

In most application domains *attribute-value based case representations* are sufficient to represent all information required to reuse cases efficiently. Such a representation consists of a set of attributes $A_1, A_2, \ldots, A_n$ where each attribute A_i is a tuple (A_{name}, A_{range}). Here, A_{name} represents the name of the attribute and A_{range} defines the set of valid values that can be assigned to that attribute. Depending on the value type (numeric or symbolic) of the attribute, A_{range} may be defined in form of an interval $[v_{min}, v_{max}]$ or in form of an enumeration $\{v_1, v_2, \ldots, v_s\}$. In principle the set of attributes might be divided into two subsets, one for the case characterization and one for the *lesson* part,

[2] A query to a CBR system can also be seen as a (partially known) case.

which describes the known solution in traditional application tasks. Further, the set of all attribute definitions, i.e. the names and the ranges, is called the *case model*. A case c is then a vector of attribute values $(a_1, a_2, \ldots, a_n)$ with $a_i \in A_i.A_{range} \cup \{undefined\}$.

An illustration of the described representation formalism is shown in Figure 2. Here, cases represent descriptions of personal computers, for example, to be used in a product recommendation system. The lesson part plays only a minor role, since a clear distinction between problems and solutions is not given in this application scenario.

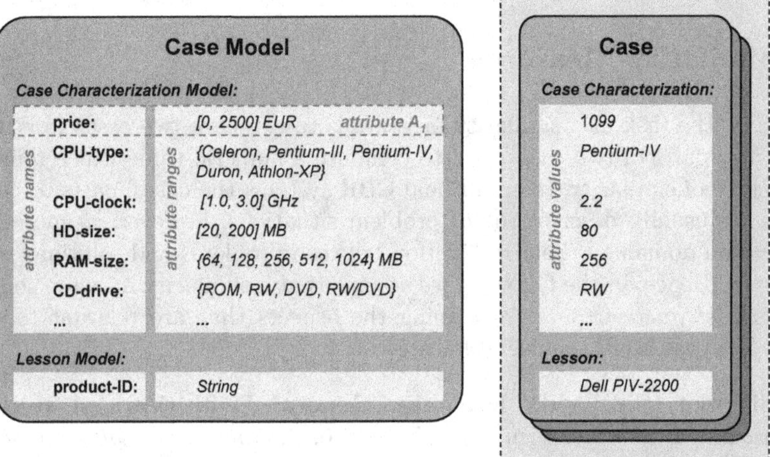

Fig. 2. Example: Attribute-Value Based Case Representation

2.2 Foundations

Although a similarity measure computes a numerical value (cf. Definition 1), usually one is not really interested in these absolute similarity values, because their interpretation is often quite difficult. One is rather interested in the partial order these values induce on the set of cases. Such a partial order can be seen as a preference relation, i.e. cases that are considered to be more similar to the query are preferred to being reused during the further processing steps.

Definition 2 (Preference Relation Induced by Similarity Measure). *Given a case characterisation $d \in \mathbb{D}$, a similarity measure Sim induces a* ***preference relation*** $\preceq_d^{Sim}$ *on the case space $\mathbb{C}$ by $c_i \preceq_d^{Sim} c_j$ iff $Sim(d, c_i) \leq Sim(d, c_j)$.*

In general, we do not assume that similarity measures necessarily have to fulfill general properties beyond Definition 1. Nevertheless, in the following we introduce two basic properties that are often fulfilled.

Definition 3 (Reflexivity). *A similarity measure is called* ***reflexive*** *if* $Sim(x, x) = 1$ *holds for all* x. *If it holds additionally* $Sim(x, y) = 1 \rightarrow x = y$, *Sim is called* ***strong reflexive.***

On the one hand, reflexivity is a very common property of similarity measures. It states that a case characterisation is maximal similar to itself. From the utility point of view, this means, a case is maximal useful with respect to its own case characterisation. On the other hand, similarity measures are usually not strong reflexive, i.e. different cases may be maximal useful regarding a given query. For example, different solution alternatives contained in different cases might be equally accurate to solve a given problem.

Definition 4 (Symmetry). *A similarity measure is called* ***symmetric,*** *if it holds* $Sim(x, y) = Sim(y, x)$ *for all* x, y. *Otherwise it is called* ***asymmetric.***

Symmetry is a property often assumed in traditional interpretations of similarity. However, in many application domains it has been emerged that an accurate utility approximation can only be achieved with asymmetric similarity measures. The reason for this is the assignment of different roles to the case characterizations being compared during utility assessment. This means, the case characterisation representing the query has another meaning than the case characterisation of the case to be rated.

2.3 The Local-Global Principle

Definition 1 still does not define how to represent a similarity measure in practice. To reduce the complexity, here one usually applies the so-called *local-global principle.* According to this principle it is possible to decompose the entire similarity computation in a local part only considering *local similarities* between single attribute values, and a global part that computes the *global similarity* for whole cases based on the local similarity assessments. Such a decomposition simplifies the modelling of similarity measures significantly and allows to define well-structured measures even for very complex case representations consisting of numerous attributes with different value types.

This approach requires the definition of a case representation consisting of attributes that are independent from each other with respect to the utility judgements. In the case that the utility depends on relations between attributes one has to introduce additional attributes — so-called *virtual attributes* — making these relations explicit. Consider the example that we want to decide whether a given rectangle is a quadrat or not by applying case-based reasoning. Here, obviously the ratio between the length and the width of the rectangle is crucial. Hence, if the original cases are described by these two attributes, an additional attribute "length-width-ratio" has to be introduced which represents this important relationship between the two attributes.

Given such a case representation, a similarity measure can be represented by the following elements:

1. *Attribute weights* define the importance of each attribute with respect to the similarity judgement,
2. *Local similarity measures.* calculate local similarity values for single attributes.
3. An *aggregation function* calculates the *global similarity* based on the attribute weights and the computed local similarity values.

With respect to the aggregation function mostly a simple weighted sum is sufficient. This leads to the following formula for calculating the global similarity Sim between a query q and a case c:

$$Sim(q, c) := \sum_{i=1}^{n} w_i \cdot sim_i(q_i, c_i)$$

Here, sim_i represents the local similarity measure for attribute a_i and q_i,c_i are the corresponding attribute values of the query and the case. In the following, we describe how local similarity measures can be represented.

2.4 Local Similarity Measures

In general, the representation of a local similarity measure strongly depends on the value type of the underlying attribute. Basically, we can distinguish between similarity measures for unordered and ordered data types. The former ones are typical for symbolic attributes while the latter ones are typical for numeric attributes. Nevertheless, also symbolic values may be associated with an order.

Similarity Tables. When dealing with unordered data types, the only feasible approach to represent local similarity measures is an explicit enumeration of all similarity values for each possible value combination. The result is a *similarity table* as illustrated in Figure 3 for the attribute 'casing" in the PC domain.

Since similarity measures are usually reflexive (cf. Definition 3), the values of the main diagonal in such a table are set to 1. When dealing with a symmetric similarity measure the upper and the lower triangular matrices are symmetric, which reduces the modelling effort. The shown table represents an asymmetric measure. For example, the similarity between the casing types "mini-tower" and "midi-tower" is different depending on which value represents the query and which the case value.

q \ c	laptop	mini-tower	midi-tower	big-tower
laptop	1.0	0.2	0.1	0.0
mini-tower	0.3	1.0	**0.9**	0.5
midi-tower	0.2	**0.7**	1.0	0.7
big-tower	0.1	0.4	0.6	1.0

Fig. 3. Similarity Table

Difference-Based Similarity Functions. For ordered types which are often also even infinite, similarity tables are not feasible, of course. Here, similarity values can be calculated based on the difference between the two values to be compared. This can be realized with *difference-based similarity functions* as illustrated in Figure 4. Here, the x-axis represents the difference between the query and the case value. For numeric types this difference can be calculated directly, for example with $\delta(q, c) = c - q$. For other ordered non-numeric types (e.g., ordered symbolic types), the distance may be inferred from the position of the values within the underlying order, i.e. one might assign an integer value to each symbolic value according its index. A difference-based similarity function then assigns every possible distance value a corresponding similarity value by considering the domain specific requirements. The function shown in Figure 4, for example, might represent the local utility function for an attribute like "price" typically occurring in product recommendation systems. The semantics of this function is that lower prices than the demanded price are acceptable and therefore lead to a similarity of 1. On the other hand, larger prices reduce the utility of a product and therefore lead to decreased similarities.

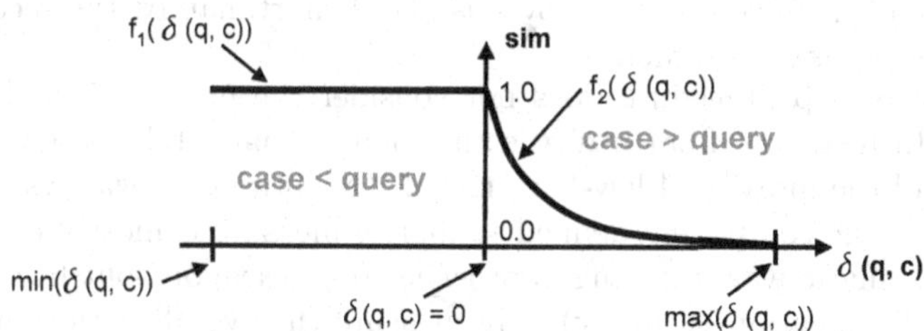

Fig. 4. Difference-Based Similarity Functions

2.5 Defining Similarity Measures

Although today available CBR tools provide comfortable graphical user interfaces for defining similarity measures, modelling similarity measures manually leads to some problems.

The similarity definition process commonly applied nowadays can be characterized as a *bottom-up* procedure. This means, the entire similarity assessment is based on the acquisition of numerous single knowledge entities about the influences on the utility function. These knowledge-entities have to be encoded separately by using suitable local similarity measures and accurate attribute weights. Because this knowledge is very specific and detailed (e.g. a local similarity measure concerns only one single aspect of the entire domain), it could also be characterised as *low-level knowledge* about the underlying utility function. Of course, to be able to acquire such general domain knowledge, at least a partial understanding of the domain is mandatory. The basic assumption of this procedure is that thorough acquisition and modelling of this low-level knowledge will lead to an accurate approximation of the complete utility function. However, in certain situations this bottom-up procedure to defining similarity measures might lead to some crucial drawbacks:

- The procedure is very time-consuming. For example, consider a symbolic attribute with 10 allowed values. This will require the definition of a similarity table with 100 entries!
- In some application domains a sufficient amount of the described low-level knowledge might be not available. Possible reasons are, for example, a poorly understood domain, or the fact that an experienced domain expert who could provide the knowledge is not available or too expensive.
- Even if an experienced domain expert is available, s/he is usually not familiar with the similarity representation formalisms of the CBR system. So, the provided knowledge may only be available in natural language. This informal knowledge then has to be translated into the formal representation formalisms by an experienced knowledge engineer who possesses the required skills which leads to additional costs.
- Due to the effort of the representation, even experienced knowledge engineers often make definition failures by mistake. Unfortunately, the recognition of such failures is very difficult.
- The bottom-up procedure does not consider the utility of whole cases directly. Instead, the final utility estimation is completely based on the ensemble of the individual low-level knowledge entities. Nowadays, the overall quality of the completely defined similarity measure is mostly not validated in a systematic way. Existing approaches (e.g. leave-one-out tests and measuring classification accuracy) only measure the overall performance of the CBR system, that is, of course, also influenced by other aspects, for example, the quality of the case data. So, one often blindly trusts the correctness of the global similarity values computed by the defined measure.
- Due to the complexity of the bottom-up procedure its application is usually restricted to the development phase of the CBR application. This means, all similarity knowledge is acquired during the development phase and is assumed to be valid during the entire lifetime of the application. However, in many domains changing requirements and/or changing environments require not only maintenance of case knowledge, but also maintenance of general knowledge [13].

- The knowledge about the actual utility of cases might not be available at all during the development phase. For example, when applying similarity measures in an e-Commerce or knowledge management scenario, the knowledge often can only be provided by the users themselves during the usage of the system. However, here the bottom-up procedure is not feasible.
- Sometimes the required knowledge about the cases' utility might already be available in a formal but quite different representation form. For example, when supporting case adaptation, the utility of cases strongly depends on the provided adaptation possibilities. Hence, to obtain an accurate similarity measure one has to transfer adaptation knowledge into the similarity measure. When using the bottom-up procedure this is a very time-consuming task. The adaptation knowledge has to be analysed manually and the knowledge considered to be relevant then has to be encoded into the similarity measure.

In the following we present an alternative approach for modelling similarity measures which tries to avoid the mentioned problems by applying machine learning techniques.

3 Learning Similarity Measures from Utility Feedback

As described in the previous section, one problem of the manual definition of similarity measures is the necessity to analyse the underlying utility functions in detail in order to determine the influences on them. Only if the different influences are known, one is able to consider them in form of appropriate weights and local similarity measures.

3.1 Utility Feedback

We have proposed an alternative approach to acquire knowledge about the only partially or informally known utility function [15,17]. The basic idea of this approach is the capability of some *similarity teacher* to give feedback about the utility of given cases with respect to concrete problem situations or queries, respectively. Here, the similarity teacher must not be able to explain the reasons why cases are more or less useful, but he has only to compare cases according their utility. This means, the utility of a case must not be expressed absolutely, but only relatively to other cases, for example, by giving statements like "case c_2 is more useful than case c_1". The similarity teacher first might analyze the result of a similarity-based retrieval, i.e. a given partial order of cases (see Figure 5). By reordering the cases according to their actual utility for a given query one obtains an additional partial order which can be characterized as a corrected retrieval result, also called *training example.*

In principle, such *utility feedback* might be provided by different types of similarity teachers depending on the concrete application scenario:

Human Domain Expert: The most obvious possibility is a *human domain expert* who posses implicit knowledge about the unknown domain specific utility function to be approximated. Due to his/her experiences, a domain expert should be able to decide which cases are more useful than others for a given problem situation.

System Users: In application scenarios where the utility of cases strongly depends on the preferences and expectations of the system's users (e.g. e-Commerce applications), also the *users* might play the role of the similarity teacher.

Software Agents: Generally, the similarity teacher has not necessarily be represented by a human being. In some application scenarios also *software agents* which are able to evaluate retrieval results automatically might be applied.

Application Environment: If the output of a CBR system is directly applied in some *application environment*, also feedback about the cases' application success or failure might lead to the required utility feedback.

In the following we assume the existence of some arbitrary similarity teacher who is able to provide utility feedback containing implicit knowledge about the unknown utility function. The objective of our approach is to extract this knowledge and to encode it in an explicit form by defining appropriate similarity measures. This can be achieved by applying machine learning techniques as described in the following.

3.2 Evaluating Similarity Measures

The foundation of our learning approach is the definition of a special *error function* E which compares retrieval results computed according to a given similarity measure with utility feedback provided by the similarity teacher (see Figure 5). This means, the basic element of the error function E has to be a measure for the distinction between two partial orders, namely a retrieval result and utility feedback with respect to a given query q. The requirement on this measure is that it should compute an error value of zero, if and only if the two partial orders are equal. Otherwise it should compute an error value greater zero representing the degree of distinction. Such an error function then can be used to evaluate the quality of a given similarity measure, since it is our goal to find a similarity measure which produces partial orders as defined by the similarity teacher. This means if we are able to find a similarity measure leading to an error value of zero, we have found an *ideal similarity measure* with respect to the given utility feedback.

In the following we introduce a possible definition of such an error function E. For better understanding, here, we introduce a simplified version of E. A more sophisticated one can be found in [17]. Before being able to compare entire partial orders, we need the following function:

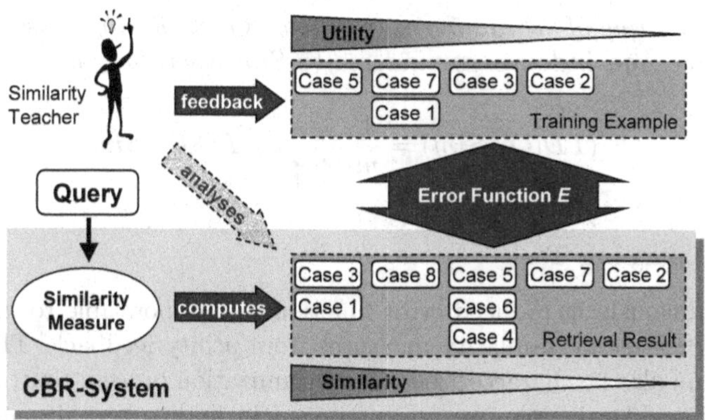

Fig. 5. Utility Feedback

Definition 5 (Elementary Feedback Function). *Let q be a query, c_i and c_j be two cases, and let Sim be a similarity measure. The function ef defined as*

$$ef(Sim,(c_i,c_j)) := \begin{cases} 1 \;\; if \;\; sgn(u(q,c_i) - u(q,c_j)) \neq sgn(Sim(q,c_i) - Sim(q,c_j)) \\ 0 \;\; otherwise \end{cases}$$

is called ***elementary feedback function*** *where u represents the informal utility feedback provided by some arbitrary similarity teacher.*

The objective of this elementary feedback function is the evaluation of a given similarity measure Sim regarding the correct ranking for a particular case pair (c_i, c_j). This function can now be used to define the mentioned measure for evaluating retrieval results:

Definition 6 (Index Error). *Consider a similarity measure Sim, a query q and utility feedback for a set of cases $\{c_1, \ldots, c_n\}$ with respect to q, also called* ***training example*** *$TE(q)$. We define the* ***index error*** *induced by Sim w.r.t. to $TE(q)$ as*

$$E_I(TE(q), Sim) = \sum_{i=1}^{n-1} \sum_{j=i+1}^{n} ef(Sim,(c_i,c_j))$$

.

The index error can be seen as a measure for the quality of a given similarity measure regarding a particular query and the corresponding training example. However, we are interested in similarity measures that supply reasonable case rankings for arbitrary queries or at least for a certain set of queries. This leads to the following extension of the index error allowing the evaluation of similarity measures with respect to a set of training examples:

Definition 7 (Average Index Error). *Consider a set of training queries $Q = \{q_1, q_2, \ldots, q_m\}$, corresponding training data $TD_u = \{TE(q_1), TE(q_2), \ldots,$*

$TE(q_m)\}$ consisting of m training examples, and a similarity measure Sim. We define the ***average index error*** *induced by Sim w.r.t. to Q as*

$$\hat{E}_I(TD(Q), Sim) = \frac{1}{m} \cdot \sum_{i=1}^{m} E_I(TE(q_i), Sim)$$

3.3 The Learning Task

With the previously introduced error function we are now able to implement a procedure for learning similarity measures from utility feedback. This learning procedure can also be characterized as an optimization process controlled by the error function. In principle, we are interested in finding an *optimal similarity measure* leading to a minimal error value, i.e. we want to find a global minimum of the error function $\hat{E}_I$ (see Figure 6). Unfortunately, in general it cannot be guaranteed that we are able to find actually a global minimum, nevertheless we are interested to minimize the error value as far as possible. When starting with some initial similarity measure $Sim_{initial}$ coupled with a corresponding error value $e_{initial}$, it should at least be possible to find a measure coupled with an error value smaller than $e_{initial}$, e.g. Sim_{supopt}. This measure then hopefully represents a better approximation of the unknown utility function.

It must be pointed out that the search space, i.e. the set of representable similarity measures, usually does not contain an ideal similarity measure. Hence, even an optimal similarity measure is mostly also coupled with an error value greater zero.

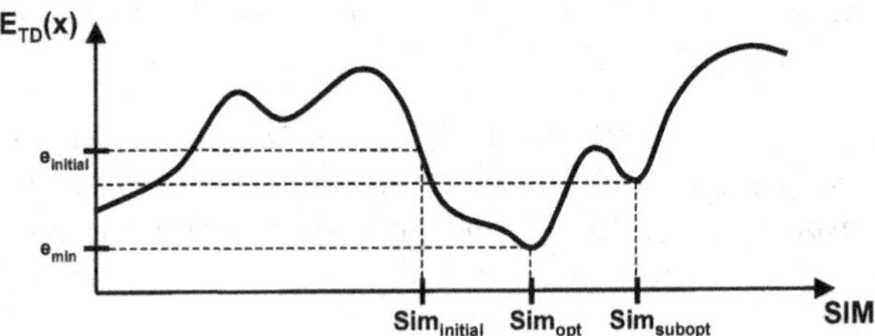

Fig. 6. Finding Minima of the Error Function

For implementing the described learning or optimization task, respectively, different methods, for example, gradient descent approaches, simulated annealing [2] or evolutionary algorithms [8,10], have already been developed. In the following section we present an approach particularly suited to learn local similarity measures that is based on evolutionary programs [10]. In contrast to traditional genetic algorithms where the entities to be optimized are encoded by using bit strings, our evolutionary algorithm operates on more sophisticated representations. We restrict the description of our approach on an overview of the most important aspects, namely the representation of individuals and the definition of appropriate genetic operators. For more details about the functionality of our learning algorithm we refer to [18,17].

4 A Genetic Algorithm for Learning Local Similarity Measures

When employing a genetic algorithm, the most important issues are the definition of an appropriate *fitness function*, an adequate representation of the entities to be optimized and the determination of corresponding genetic operators. The average index error $\hat{E}_I$ introduced in Definition 7 already represents the required fitness function. In this section we show how local similarity measures can be represented so that a genetic algorithm is able to handle them easily. Further, we introduce corresponding genetic operators needed to realize an evolutionary process.

4.1 Representation of Individuals

Concerning the representation of local similarity measures as individuals of an evolutionary process we presume the representation formalisms introduced in Section 2.4, i.e. difference-based similarity functions and similarity tables.

Representing Difference-Based Similarity Functions. Consider some difference-based similarity function Sim_A used as local similarity measure for a numeric attribute A. Since Sim_A may be continuous in its value range $[min(A_{range}) - max(A_{range}), max(A_{range}) - min(A_{range})]$ it is generally difficult to describe it exactly with a fixed set of parameters. Thus, we employ an approximation based on a number of sampling points to describe arbitrary functions:

Definition 8 (Similarity Function Individual, Similarity Vector). *An individual I representing a similarity function Sim_A for the numeric attribute A is coded as a vector V_A^I of fixed size s. The elements of that* ***similarity vector*** *are interpreted as sampling points of Sim_A, between which the similarity function is linearly interpolated. Accordingly, it holds for all $i \in \{1, \ldots, s\}$: $v_i^I = (V_A^I)_i \in [0, 1]$.*

The number of sampling points s may be chosen due to the demands of the application domain: The more elements V_A^I contains, the more accurate the approximation of the corresponding similarity function, but on the other hand, the higher the computational effort required for optimization. Depending on the characteristics of the application domain and the particular attribute, different strategies for distributing sampling points over the value range of the similarity function might be promising:

Uniform Sampling: The simplest strategy is to distribute sampling points equidistantly over the entire value range (see Figure 7a).

Center-Focused Sampling: However, when analyzing the structure of difference-based similarity functions in more detail, it becomes clear that different inputs of the functions will usually occur with different probabilities. While the maximal and minimal inputs, i.e. the values $d_{min} = (min(A_{range}) - max(A_{range}))$ and $d_{max} = (max(A_{range}) - min(A_{range}))$, can only occur for one combination of query and case values, inputs corresponding to small differences can be generated by various of such combinations. Thus, case data and corresponding training data usually provides much more information about the influences of small differences, compared with the information available about extreme differences. Another aspect is that changes in similarity are usually more important for small value differences, since greater differences usually correspond to very small similarity values. In order to consider these facts during learning, it might be useful to use more sampling points around the "center" of the difference-based similarity function like illustrated in Figure 7b.

Dynamic Sampling: While the center-focused approach is more a heuristics, it is also possible to analyse the training data in order to determine an optimal distribution for the sampling points [7]. Then, areas where a lot of training information is available might be covered with more sampling points than areas for which no respective information is contained in the training data.

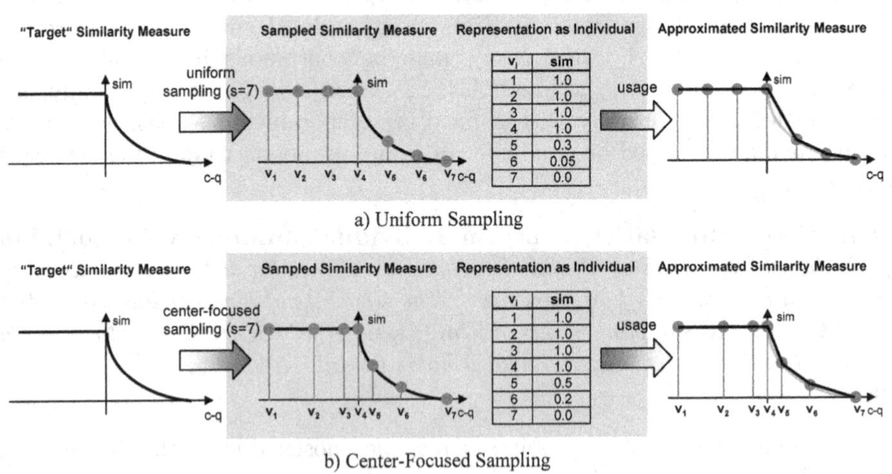

Fig. 7. Representing Similarity Functions as Individuals

Representing Similarity Tables. Similarity tables, as the second type of local similarity measures of concern, are represented as matrices of floating point numbers within the interval $[0, 1]$:

Definition 9 (Similarity Table Individual, Similarity Matrix). *An individual I representing a similarity table for a symbolic attribute A with a list of allowed values $A_{range} = (d_1, d_2, \ldots, d_n)$ is a $n \times n$-matrix M_A^I with entries $m_{ij}^I = (M_A^I)_{ij} \in [0,1]$ for all $i, j \in \{1, \ldots, n\}$.*

This definition corresponds to the representation of similarity tables, i.e. the original representation of this type of local similarity measures is directly used by the genetic algorithm. So, the definition presented here is only required for introducing the necessary notation.

4.2 Genetic Operators

Another important issue is the definition of accurate genetic operators used to perform crossover and mutation operations. When deciding not to use bit strings, but other data structures for representing individuals, the genetic operators have to consider the particularly used genome representation. Therefore, in this section also some exemplary genetic operators for the previously introduced genome representation are presented.

The operators we use for learning of local similarity measures are different from classical ones since they operate on a different genome representation. However, because of underlying similarities, we divide them also into the two standard groups: mutation and crossover operators.

Crossover Operators for Similarity Vectors and Matrices. Applying crossover operators on the data structures used for representing local similarity measures, a new individual in the form of a similarity vector or matrix is created using elements of its parents. Though there are variations of crossover operators described that exploit an arbitrary number of parents [10], we rely on the traditional approach using exactly two parental individuals, I_1 and I_2.

- *Simple crossover* is defined in the traditional way as used for bit string representations: A split point for the particular similarity vector or matrix is chosen. The new individual is assembled by using the first part of parent I_1's similarity vector or matrix and the second part of parent I_2's.
- *Arbitrary crossover* represents a kind of multi-split-point crossover with a random number of split points. Here, for each component of the offspring individual it is decided randomly whether to use the corresponding vector or matrix element from parent I_1 or I_2.
- *Arithmetical crossover* is defined as the linear combination of both parent similarity vectors or matrices. In the case of similarity matrices the offspring is generated according to: $(M_A^{I_{new}})_{ij} = m_{ij}^{I_{new}}$ with $m_{ij}^{I_{new}} = \frac{1}{2} m_{ij}^{I_1} + \frac{1}{2} m_{ij}^{I_2}$ for all $i, j \in \{1, \ldots, d\}$.
- *Line/column crossover* is employed for similarity tables, i.e. for symbolic attributes, only. Lines and columns in a similarity matrix contain coherent information, since their similarity entries refer to the same query or case value, respectively. Therefore, cutting a line/column by simple or arbitrary

crossover may lead to less valuable lines/columns for the offspring individual. We define line crossover as follows: For each line $i \in \{1, \ldots, n\}$ we randomly determine individual I_1 or I_2 to be the parent individual I_P for that line. Then it holds $m_{ij}^{I_{new}} = m_{ij}^{I_P}$ for all $j \in \{1, \ldots, n\}$. Column crossover is defined accordingly.

For each of the described operators a particular probability value has to be specified. When performing crossover, one of the described operators is then selected according to this probability.

Mutation Operators for Similarity Vectors and Matrices. Operators of this class are the same for both kinds of local similarity measures we are dealing with. They change one or more values of a similarity vector V_A^I or matrix M_A^I according to the respective mutation rule. Doing so, the constraint that every new value has to lie within the interval $[0, 1]$ is met. The second constraint that needs to be considered concerns the reflexivity of local similarity measures (cf. Definition 3). As a consequence, the medial sampling point of a similarity vector should be 1.0 as well as the elements m_{ii}^I of a similarity matrix for all $i \in \{1, \ldots, n\}$. Since any matrix can be understood as a vector, we describe the functionality of our mutation operators for similarity vectors only:

- *Simple mutation*: If $V_A^I = (v_1^I, \ldots, v_s^I)$ is a similarity vector individual, then each element v_i^I has the same probability of undergoing a mutation. The result of a single application of this operator is a changed similarity vector $(v_1^I, \ldots, \hat{v}_j^I, \ldots, v_s)$, with $1 \leq j \leq s$ and $\hat{v}_j^I$ chosen randomly from $[0, 1]$.
- *Multivariate non-uniform mutation* applies the simple mutation to several elements of V_a^I. Moreover, the alterations introduced to an element of that vector, become smaller as the age of the population is increasing. The new value for v_j^I is computed after $\hat{v}_j^I = v_j^I \pm (1 - r^{(1-\frac{t}{T})^2})$, where t is the current age of the population at hand, T its maximal age, and r a random number from $[0, 1]$. Hence, this property makes the operator search the space more uniformly at early stages of the evolutional process (when t is small) and rather locally at later times. The sign $\pm$ indicates, that the alteration is either additive or subtractive. The decision about that is made randomly as well.
- *In-/decreasing mutation* represents a specialisation of the previous operator. Sometimes it is helpful to modify a number of neighbouring sampling points uniformly. The operator for in-/decreasing mutation randomly picks two sampling points v_j^I and v_k^I and increases or decreases the values for all v_i^I with $j \leq i \leq k$ by a fixed increment.

As for crossover operators, mutation operators are applied according to some probability to be specified a priori.

With the described representation and genetic operators together with the error function introduced in Definition 7 we are now able to implement a genetic algorithm for learning local similarity measures. For more details about the general functionality of genetic algorithms see [8,10].

For the other important part of a global similarity measure, namely the attribute weights, it is also possible to define a corresponding genetic algorithm. However, here other learning strategies can usually be applied more efficiently, for example, gradient descent algorithms [15,16,17].

5 Experimental Evalutation

In this section we give a short summary of an experimental evaluation that demonstrates the capabilities of our learning approach in two different application scenarios. A more detailed description is given by [17]. In both scenarios a similarity measure is required to approximate an a-priori unknown utility function. However, the two scenarios clearly differ in the aspects that determine the utility of cases.

5.1 Learning Customer Preferences

In our first evaluation scenario we consider a CBR system used to recommend appropriate used cars with respect to a given requirement specification represented by the query. Here, we assume that the cars cannot be customized, i.e. no case adaptation is performed after the similarity-based retrieval. The case representation we used for our experiment consists of 8 attributes (4 symbolic, 4 numeric) describing important properties of the cars, like "price" or "engine power". Concerning the similarity measure, this results in 4 similarity tables, 4 difference-based similarity functions and 8 attribute weights to be optimized by applying our learning approach.

Since no real customers were available, we have applied a simulation approach in order to obtain the required utility feedback. The foundation of this simulation is an additional similarity measure representing virtual preferences of some class of customers, so to speak the target measure to be learnt by the learning algorithm. With this additional similarity measure Sim_U we were able to generate utility feedback like shown in Figure 8. In order to obtain a single training example, the following steps are performed automatically:

1. Generating a random query q.
2. Retrieving the 10 most similar cases with respect to q by using an initial similarity measure Sim_I.
3. Recalculating the similarity of the 10 retrieved cases by using Sim_U.
4. Selecting the 3 most similar cases with respect to Sim_U, however by incorporating some noise.

The idea of step four is a realistic simulation of the behaviour of customers. One the one hand, customers will usually not be willed to give feedback about the entire retrieval result, however to get feedback about only three cases should be realistic. Further we have introduced noise in order to simulate inconsistent behaviour of customers. Hence, the finally obtained training example does not always come up to the target measure Sim_U, but may contain minor or major deviations according to some probability values ρ_i.

By repeating the described procedure we were able to generate an arbitrary number of training examples to be used by the learning algorithm. In our experiment we have used a genetic algorithm to learn attribute weights and local similarity measures (cf. Section 4).

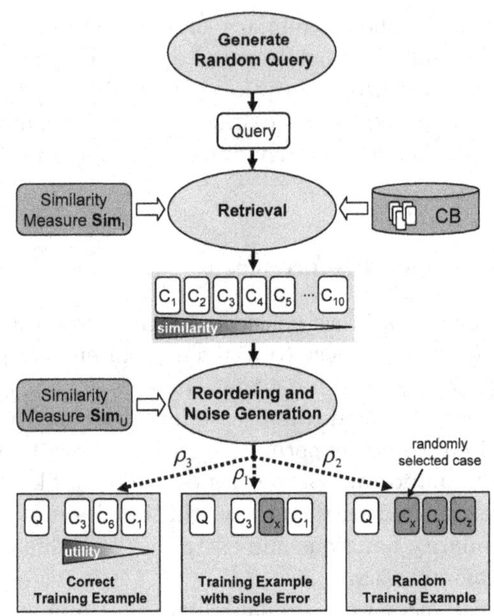

Fig. 8. Learning Customer Preferences: Generation of Training Examples

In order to be able to measure the quality of the learned similarity measure we have generated 200 additional noise free test examples. For a given learned similarity measure Sim_L we then counted the percentage of retrievals based on the 200 queries of the test examples where

- the most similar case was also the most useful one (*1-in-1*)
- the most similar case was at least among the ten most useful ones (*1-in-10*)

according to the noise free utility feedback of the test examples. In order to get an impression of the amount of training data required to get meaningful results, we have started the learning algorithm for an increasing number of training examples. Further, we have repeated the entire experiment at least 3 times in order to achieve average values, even though the number of repetitions is to small for obtaining statistically significant results, of course.

The results of the described experiment are illustrated in Figure 9. Generally, one observes clear improvements in both quality measures at least when

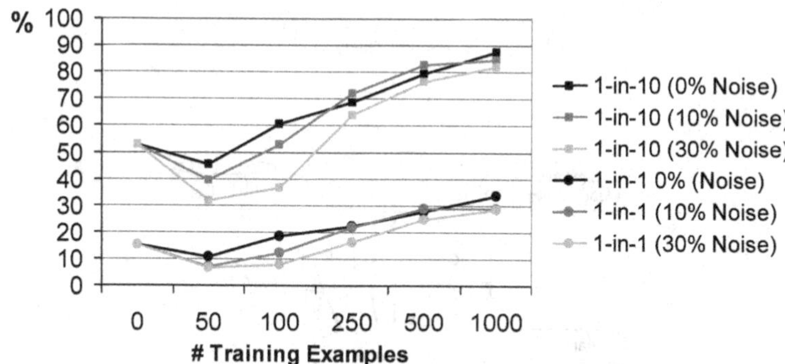

Fig. 9. Learning Customer Preferences: Results

providing more then 100 training examples. When using less examples we notice a decrease of retrieval quality which can be explained by overfitting the provided training data, so that the quality of the learned similarity measure is bad with respect to independent test examples. Further, we see that noise seems to have a significant impact on learning only when providing less then 250 training examples.

5.2 Learning to Retrieve Adaptable Cases

For our second experiment we suppose again a product recommendation system, this time used for recommending optimal configurations of personal computers. Since PCs can easily be customized by adding or replacing components, here we assume a CBR system that provides adaptation functionality. This means, the case base contains a set of base PC configurations[3] which can be modified by applying certain adaptation rules [5]. In our example domain PCs are described by 11 attributes (5 numeric and 6 symbolic) and can be adapted by applying 15 more and less complex adaptation rules.

When providing adaptation functionality, optimal retrieval results can only be achieved if the similarity measure considers the adaptation possibilities [16, 14,11]. Since the utility of a case might change clearly after being adapted, it is not sufficient to estimate the direct utility of cases for a given query. However, to define a similarity measure that estimates the utility of a case that can be achieved by adaptation, one has to analyze the available adaptation knowledge in detail. Further, the knowledge has to be transferred into the similarity measure by using the given knowledge representations. Because this is a very complex and time consuming process, we propose to automate it by applying our learning framework.

Therefore, we assume that a similarity measure Sim_U which estimates the direct utility of cases without considering adaptation possibilities is already given. Usually such a measure can be defined much more easily, or it might also be

[3] The case base used for the experiment contained 15 base configurations

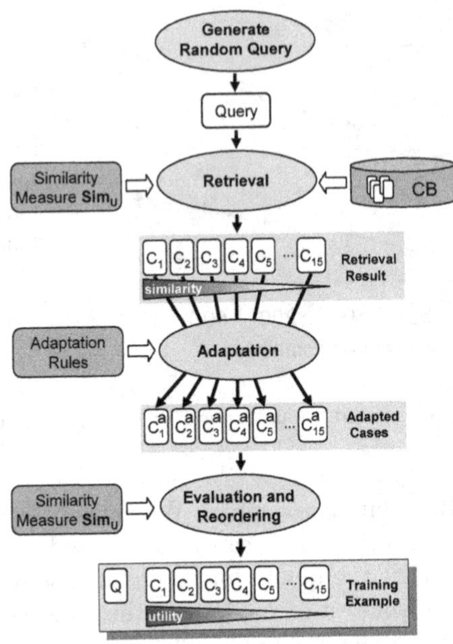

Fig. 10. Learning to retrieve Adaptable Cases: Generation of Training Examples

learned, e.g. like described in Section 5.1. Then, we are able to generate utility feedback like illustrated in Figure 10:

1. Generating a random query q.
2. Retrieving the 10 most similar cases with respect to q by using the similarity measure Sim_U or some other initial measure.
3. Adapting the retrieved cases by applying the adaptation rules.
4. Recalculating the similarity of the 10 adapted cases by using Sim_U.
5. Constructing a training example by reordering the 10 original cases.

This feedback can be used to learn a new similarity measure Sim_A which can be seen as an optimized version of Sim_U since it tries to approximate the utility of cases under consideration of adaptation possibilities. The achieved results are illustrated in Figure 11. The results shown here represent average values obtained during 10 repetitions of the experiment. The experimental settings were similar to the experiment described in Section 5.1. Again we have applied our learning algorithm by using increasing number of training examples. To measure the quality of the learned similarity measures we have determined the corresponding retrieval results for 200 independent test queries. Instead of the *1-in-10* quality measure, here, we have chosen the analogous *1-in-3* version. The idea of this measure is the assumption that it is computational feasible to adapt the 3 most similar cases after retrieval in order to select the best resulting case.

For both selected quality measures we notice a clear improvement of the retrieval quality achieved with the optimized similarity measure when using more

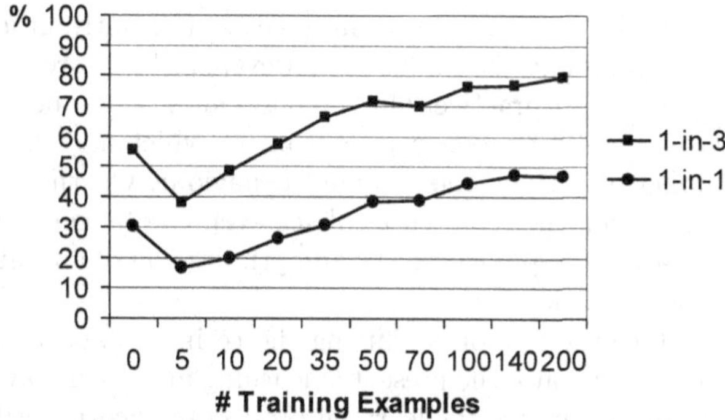

Fig. 11. Learning to retrieve Adaptable Cases: Results

then 20-35 training examples. In this experiment less training examples were necessary to avoid overfitting because each training example contained more knowledge (feedback about 10 cases instead 3 cases in the previous scenario).

6 Conclusion

We have discussed that the output of knowledge-based systems cannot always simply be judged as correct or incorrect as one would expect from a problem-solving system. In many application domains the output of such systems rather has to be interpreted by considering certain utility functions. This means the output might be more or less useful for solving a problem or for satisfying the users' demands. In order to produce maximal useful outputs, a knowledge-based system should be able to estimate the utility of a possible output a-priori.

In CBR systems, for example, the utility of available knowledge is approximated by employing similarity measures. The more domain specific knowledge one is able to encode into a similarity measure the higher should be the probability that useful knowledge is selected. According to the local-global principle, particular local similarity measures can be used to encode much domain specific knowledge about the utility function to be approximated. However, the definition of such knowledge-intensive similarity measures is a complex and time consuming process.

In this article we have proposed to facilitate the definition of knowledge-intensive similarity measures by applying a machine learning approach. The approach is based on feedback about the actual utility of cases provided by some similarity teacher. This feedback enables us to evaluate the quality of given similarity measures and can be used to guide an search process with the goal to find an optimal similarity measure. We have described a genetic algorithm for realising this search or optimization process, respectively. In order to show the capabilities of our learning approach, finally we have presented the results of two different evaluation experiments. Here, we have seen that the quality of an initially given similarity measure can be improved significantly, at least if reasonable amount of training data is available. However, in both presented scenarios this should be possible in practice. On the one hand, a product recommendation system is usually used by numerous customers, which may provide feedback explicitly or implicitly, e.g. by their buying behaviour. On the other hand, in our adaptation scenario, the required feedback even can be generated automatically. Here our learning approach allows an optimization of an initial similarity measure "on a mouse click".

In order to reduce the risk of overfitting the training data, and thus reduce the amount of necessary data, the presented learning approach may be extended. Several possibilities to improve the learning process by incorporating additional background knowledge are presented in [7].

References

1. A. Aamodt and E. Plaza. Case-based reasoning: Foundational Issues, Methodological Variations, and System Approaches. *AI Communications*, 7(1):39–59, 1994.
2. E. Aarts and J. Korst. *Simulated Annealing and Boltzmann Machines.* John Wiley & Sons, 1989.
3. R. Bergmann, M. Michael Richter, S. Schmitt, A. Stahl, and I. Vollrath. Utility-Oriented Matching: A New Research Direction for Case-Based Reasoning. In *Professionelles Wissensmanagement: Erfahrungen und Visionen. Proceedings of the 1st Conference on Professional Knowledge Management.* Shaker, 2001.
4. R. Bergmann, S. Schmitt, and A. Stahl. *E-Commerce and Intelligent Methods*, chapter Intelligent Customer Support for Product Selection with Case-Based Reasoning. Physica-Verlag, 2002.
5. R. Bergmann, W. Wilke, I. Vollrath, and S. Wess. Integrating General Knowledge with Object-Oriented Case Representation and Reasoning. In *Proceedings of the 4th German Workshop on Case-Based Reasoning (GWCBR'96)*, 1996.
6. R. Burke. The Wasabi Personal Shopper: A Case-Based Recommender System. In *Proceedings of the 11th International Conference on Innovative Applications of Artificial Intelligence (IAAI'99)*, 1999.
7. T. Gabel. Learning Similarity Measures: Strategies to Enhance the Optimisation Process. Master thesis, Kaiserslautern University of Technology, 2003.
8. J.H. Holland. *Adaptation in Natural and Artificial Systems.* The University of Michigan Press, 1975.
9. M. Lenz, B. Bartsch-Spörl, H.D. Burkhard, and S. Wess, editors. *Case-Based Reasoning Technology: From Foundations to Applications.* LNAI: State of the Art. Springer, 1998.
10. Z. Michalewicz. *Genetic Algorithms + Data Structures = Evolution Programs.* Springer, 1996.
11. M.M. Richter. Learning Similarities for Informally Defined Objects. In R. Kühn, R. Menzel, W. Menzel, U. Ratsch, M.M. Richter, and I.-O. Stamatescu, editors, *Adaptivity and Learning.* Springer, 2003.
12. C. J. van Rijsbergen. *Information Retrieval.* Butterworths & Co, 1975.
13. T. Roth-Berghofer. *Knowledge Maintenance of Case-Based Reasoning Systems.* Ph.D. Thesis, University of Kaiserslautern, 2002.

14. B. Smyth and M. T. Keane. Retrieving Adaptable Cases: The Role of Adaptation Knowledge in Case Retrieval. In *Proceedings of the 1st European Workshop on Case-Based Reasoning (EWCBR'93)*. Springer, 1993.
15. A. Stahl. Learning Feature Weights from Case Order Feedback. In *Proceedings of the 4th International Conference on Case-Based Reasoning (ICCBR'2001)*. Springer, 2001.
16. A. Stahl. Defining Similarity Measures: Top-Down vs. Bottom-Up. In *Proceedings of the 6th European Conference on Case-Based Reasoning (ECCBR'2002)*. Springer, 2002.
17. A. Stahl. *Learning of Knowledge-Intensive Similarity Measures in Case-Based Reasoning.* Ph.D. thesis, Technical University of Kaiserslautern, 2003.
18. A. Stahl and T. Gabel. Using Evolution Programs to Learn Local Similarity Measures. In *Proceedings of the 5th International Conference on Case-Based Reasoning (ICCBR'2003)*. Springer, 2003.
19. W. Wilke, M. Lenz, and S. Wess. *Case-Based Reasoning Technology: From Foundations to Applications*, chapter Case-Based Reasoning and Electronic Commerce. Lecture Notes on AI: State of the Art. Springer, 1998.

Knowledge Sharing in Agile Software Teams

Thomas Chau and Frank Maurer

University of Calgary,
Department of Computer Science
Calgary, Alberta, Canada T2N 1N4
{chauth,maurer}@cpsc.ucalgary.ca

Abstract. Traditionally, software development teams follow Tayloristic approaches favoring division of labor and, hence, the use of role-based teams. Role-based teams require the transfer of knowledge from one stage of the development process to the next. As multiple stages are involved, the problem of miscommunication due to indirect and long communication path is amplified. Agile development teams address this problem by using cross-functional teams that encourages direct communication and reduces the likelihood of miscommunication. Agile approaches usually require team members to be co-located and only facilitate intra-team learning. To overcome the restriction in co-location and support organizational inter-team learning while supporting the social context critical to the sharing of tacit knowledge is the focus of this paper. We also highlight that humans are good at making sense of incomplete and approximative information.

1 Introduction

Software development is a collaborative process that needs to bring together domain expertise with technological skills and process knowledge. Traditional software development approaches organize the required knowledge sharing based on different roles following a Tayloristic mindset: people involved in the development process are assigned to specific roles (e.g. business analyst, software architect, lead designer, programmer, tester) that are associated with specific stages in the development process (requirements analysis, high-level design, low level design, coding, testing). Hand-offs between each of the stages are primarily document based: one role produces a document (e.g. a requirements specification, design documents, source code, test plans) and hands it off to the people responsible for the next stage in the development process. Moving from one stage to the next often requires sign-offs from the people involved. After sign-off, documents are supposed to be set in stone (e.g. "the design is completed") and changes are seen as a problem: feature creep, requirements churn etc. Changes are handled as the exception to the rule that need to get formal approval by change control boards. Changes are also seen as a factor that dramatically (i.e. up to two orders of magnitude) increases development costs.

Tayloristic processes strive to accomplish an idealistic goal: Documents are supposed to be complete, consistent and unambiguous. Unfortunately, they never are: the information contained in the documents only approximates what is needed by the people handling the next stage Information is lost in each transfer from one head to

W. Lenski (Ed.): Logic versus Approximation, LNCS 3075, pp. 173–183, 2004.

the next. And knowledge that does not reach the person who actually writes the code will result in a software system that does not meet all customer needs – resulting in low customer satisfaction and unsuccessful projects. A simply model illustrates this knowledge loss over longer communication chains (see Figure 1).

Customer -> Analyst -> Architect -> Designer -> Chief Programmer -> Coder	
10% communication error:	59% of information gets to coder
5% communication error:	77% of information gets to coder

Fig. 1. Knowledge loss in Tayloristic processes

Assuming only a mere 5% of relevant information is lost in each transfer between each of the stages, nearly a quarter of the information does not reach the coder (who has to encode the domain knowledge into software) in a Tayloristic development process[1]. This gets worse if more then 5% are lost in each stage. Clearly, software engineering approaches are trying to reduce this information loss by spending effort on error correction: the main purpose of reviews and inspection is to reduce errors in all kinds of documents. While there is lots of empirical evidence showing that effort spent on inspections and reviews pays of in reduced miscommunication, nothing indicates that these processes are able to come close to a 0% error. In fact, the CHAOS report published by the Standish group shows that about three quarters of all software development projects either fail completely or are challenged. One of the main reasons for this is that the software delivered does not meet customer needs.

Another problem resulting from the long communication chains in Tayloristic software organizations is a tendency to over-document. Shannon's information theory indicates that information is only useful when it is new for the receiver of the information: providing a known fact to somebody is old news and boring. In fact, if this is done in a document, it makes the task to find relevant gems of information more difficult and, hence, increases knowledge transfer costs. People involved in the early stages of software development do not (and can not) know what information is already known to the coders. Relevance of information is completely subjective in the sense that it depends on the current knowledge of the information receiver. Based on experience, analysts and designers know that incomplete information will result in implementations that do not meet the customer's expectations. To be on the safe side and avoid problems with incomplete specifications and designs, analysts and designers tend to over-document: they answer questions with their documentation that might not even be asked by the coder.

A simple way to reduce the information loss on one hand while focusing communication on relevant information is reducing the length of the communication chain (see Figure 2).

Customer -> Developer	
10% communication error:	90% of information gets to coder
5% communication error:	95% of information gets to coder

Fig. 2. Knowledge loss in direct communication

[1] Information preserved = $(100\% - \text{error rate})^{\#\text{ of transfer}}$ (e.g. $59\% = (100-10\%)^5$)

Agile software processes like Extreme Programming [2], Scrum [4] and others [1, 3, 5, 6], rely on direct face-to-face communication between customers and developers for knowledge sharing. This reduces the information loss due to long communication chains as well as making sure that only questions that the developer (who writes the code) has are answered.

Transferring and sharing required knowledge in a team is a difficult task that, in the past, was tackled by introducing rigorous processes and more and more structured and formalized representations. While there are merits to that approach, the recent trend towards agile software processes focuses on less formal, more fuzzy style. It replaces "logical" representations by approximations - approximations that are "good enough" for humans to proceed with development but rely on face-to-face sharing of tacit knowledge to actually do so.

In this paper, we will describe how agile teams share knowledge and discuss benefits and shortcomings of these approaches. We will then present a lightweight knowledge management framework that overcomes some of the problems and explain how lightweight knowledge management can be integrated with more structured knowledge. We will also review some existing tool support for agile practices from the perspective of lightweight knowledge management.

2 Knowledge Sharing Support in Agile Processes

In agile processes, knowledge sharing is encouraged by several practices: release and iteration planning, pair programming and pair rotation, on-site customers in case of XP [2], daily Scrum meeting, cross-functional teams, and project retrospectives in Scrum [4].

Release and iteration planning are used to share knowledge on system requirements and the business domain between the on-site customers and the developers. In a release planning meeting arranged at the beginning of a project, the project timeline is broken down into small development iterations and releases. At the beginning of an iteration (short time-boxed develop efforts that run usually two to six weeks), the development team and the customer representatives discuss what should be done in the next few weeks. The discussions refine the initial requirements to a level that the development team is able to estimate the development effort for each feature. Developers break each feature into tasks and provide the customers with estimates of effort needed to complete each feature. Based on the developers' estimation, the amount of work hours available in the upcoming iteration and the velocity (the percentage spent on development task in relation to total work) from the previous iteration, prediction can be made as to whether the developers can complete the features proposed by the customers. If not, the developers are to renegotiate the set of features with the customers. Further requirement details are discussed with on-site customer representatives while a developer actually works on the implementation of a feature. The close interaction between developers and on-site customer representatives usually lead to increased trust and a better understanding. This direct feedback loop allows a developer to create a good approximation of the requirements in his head faster than document-centric information exchange. Quickly developed software can be demonstrated immediately to the customer representative and allows her to directly catch misunderstandings.

Pair programming involves two developers working in front of a single computer designing, coding, and testing the software together. It is a very social process characterized by informal and spontaneous communications. During a pair programming session, knowledge of various kinds, some explicit but mostly tacit, is shared between the pair. This includes task-related knowledge, contextual knowledge, and social resources. Examples of task-related knowledge include system knowledge, coding convention, design practices, technology knowledge and tool usage tricks. Contextual knowledge is knowledge by which facts are interpreted and used. For instance, knowing from past experiences or "war stories" when to or when not to use a particular design pattern in different coding scenarios. Examples of social resources include personal contacts and referrals. Developers tend not to document these types of knowledge for many reasons, such as being overburdened with other tasks or they deem what they know to be irrelevant or of no interest to others. Such knowledge is often only uncovered via informal and casual conversation [7]. For this reason, the social nature of pair programming made it a great facilitator for eliciting and sharing tacit knowledge. To ensure knowledge shared among a pair is accessible to the entire team, XP recommends pairs be rotated from time to time. As a side effect of tapping tacit knowledge, the social nature of pair programming helps to create and strengthen networks of personal relationships within a team, and nurture an environment of trust, reciprocity, shared norms and values. These are critical to sustain an ongoing culture of knowledge sharing.

While pair programming sessions facilitate communication within a pair, daily Scrum meetings facilitate communication among the entire team. During a daily Scrum meeting, team members report their work progress since the last meeting; state their goals for the day; and voice problems related to their tasks or suggestions to their colleagues' tasks. Such meetings provide visibility of one's work to the rest of the team; raise everyone's awareness of who has worked on or is knowledgeable about specific parts of the system; and encourage communications among team members who may not talk to each other regularly. Team members learn whom to contact when they work on parts of the system that they are unfamiliar with.

To reduce the communication cost among the various roles, such as business analysts, developers, and testers, who are involved in software development, agile methods recommend the use of cross-functional teams instead of role-based teams. A role-based team contains only members of the same role. In contrast, a cross-functional team draws together individuals of all defined roles. Experiences indicate that cross-functional teams facilitate better collaboration and knowledge sharing which lead to reduced product development time [8].

Continuous learning is supported by some agile methods in the form of project retrospectives. Retrospectives are in essence post-mortem reviews on what happened during development except that they are conducted not only at the end of a project but also during the project. Retrospectives facilitate the identification of any success factors and obstacles of the current management and development process. In cases where team members face obstacles of the current process, such as lengthy stand-up meetings, retrospectives provide the opportunity for these issues to be raised, discussed, and dealt with during the project rather than at the end of project.

3 Limitations

The above knowledge sharing practices are all team-oriented and rely on social interactions. Although the social nature of the practices made them great at tapping tacit knowledge and in fostering the creation and reinforcement of relationship networks within a team, there are inherent limitations in these practices. In their original form, all the above practices rely on face-to-face communication which restricts the use of them to co-located and small teams, usually with less than twenty people [2]. Unfortunately, for many reasons, it is sometimes impossible to co-locate an entire team and sole reliance on informal knowledge sharing will present challenges. Hence, distributed teams and inter-team knowledge sharing are issues that agile methods practitioners must deal with.

Besides the co-location constraint, the above practices only facilitate intra-team but not inter-team learning within an organization. A common attempt to address this organizational learning issue is to transfer workers from one work team to another. However, this is challenging due to cost and is slow due to time constraints.

Informal training approaches like pair programming and pair rotation are not problem-free either. Training content may vary, or conflict across different pairs. Getting two people to work cooperatively as a pair is also often an extremely tricky task. One may argue that pair programming constantly reduces the productivity of the experts as they need to train novices all the time and formal training is therefore less expensive. It should be possible to combine pair programming with a training infrastructure to gain the benefits of both approaches.

4 MASE

To overcome the above limitations, there exist various tools ranging from those that support real-time collaboration, such as Microsoft's Messenger and NetMeeting, to those that support asynchronous communication and coordination, such as e-mail and newsgroups.

While real-time collaboration tools like NetMeeting facilitate the social interaction necessary for sharing tacit knowledge, their usage is limited only to team members working at the same time. Assuming normal work hours, this is nearly impossible for teams with members working in different time zones. Likewise, tools such as e-mail and newsgroups support only asynchronous communication and collaboration.

However, the fact that "people move continually and effortlessly between different styles of collaboration: across time, across place, and so on" [9] demands tool support that can accommodate more than one collaboration style like:

- Co-located and distributed team;
- Retrieval and use of structured and unstructured information content, and;
- Synchronous and asynchronous activities.

In addition, for such a tool to be useful to agile development teams, it needs to:

- Support the social context critical to nurturing a knowledge sharing environment – providing information needs to be as easy as accessing it;

- Facilitate organizational learning, and;
- Support specific agile practices.

We will now illustrate how our proposed lightweight knowledge management platform, MASE, is able to achieve these goals and address issues stated above.

4.1 Support for Co-located and Distributed Teams

MASE is a web-based collaboration and knowledge sharing tool for agile teams. Web technology makes the tool accessible anytime anywhere by users with a web browser in their computing environment. The tool does not distinguish users working at the same place from those who work at different places. Hence, MASE is capable of supporting collaboration for both co-located and distributed teams.

4.2 Support for Unstructured and Structured Information Content

MASE allows combining the sharing of unstructured as well as structured information. Further, it makes writing (providing information) nearly as easy as reading (accessing information). Unstructured information usually consists of text and graphics. Structured information is stored in a database and, thus, must follow a schema. To support unstructured information content, the user interface of MASE is developed based on Wiki technology [10]. Wiki enable any users to access, browse, create, structure, and update any web pages in real-time using a web browser only. Each of these web pages, known as a wiki page, acts like an electronic bulletin board discussion topic with a unique name. Users use the Wiki markup language to create Wiki pages. Wiki markup is very simple (much simpler than HTML): a list of all Wiki markup commands including examples fits onto a single page.

One may argue that wiki pages are no different from any other traditional documents and will suffer from the same maintenance problems. Wiki technology mitigates this risk by automatically creating links from a wiki page to particular topics pages if the names of those topics are mentioned in that page. This helps minimize the users' effort in maintaining the relationships among the content in different wiki pages and enhance knowledge discovery. These benefits, however, are only maximized if users adhere to the same terminology when contributing content to wiki pages.

Information content in a MASE wiki page is all free-formatted text. This is not the case when users try to update a typical web page. Typically, the information content on the web page that users see is embedded among presentation information like HTML markup elements. For the users to edit such a web page, they need to spend the extra effort to first extract the information content then begin the actual editing of the content. Sometimes, this additional effort is so time-consuming that the users often give up on editing the content, thus causing knowledge content to degenerate over time. The fact that information content of web pages in MASE is in free-formatted text facilitates efficient collaboration between knowledge contributors and readers.

To support structured information content, MASE achieves this through its library of plug-ins that store specific data in a database. A MASE plug-in is usually presented as an input form that allows users to submit information or a table that displays information retrieved from a database in a structured fashion. Users can include a MASE plug-in in any wiki page simply by referencing its name. Currently, the MASE library of plug-ins includes those that are specific to agile development teams and generic team-oriented collaboration tools like rating a specific wiki page supporting collaborative filtering. The fact that any content on any wiki pages are modifiable and that any plug-ins can be included in any wiki pages give the users the flexibility to control how structured or unstructured they want their team memory to be.

Storing information does not guarantee that others can find it. And retrieving information from a repository that combines structured and unstructured data usually requires users to learn two different query mechanisms. To overcome the resulting usability problems, MASE provides full-text searching capabilities on any unstructured and structured content.

4.3 Support for Personal Portals

When a team member first logs into MASE, she is automatically provided a portal - an individual information space. She can store in her portal any content, in either structured or unstructured format. The content may be relevant only to her or to some other team members; it may not even be related to the project or task at hand. The key idea is that a team member has complete control over the type of and the granularity of the information content that she wants to see.

4.4 Support for Asynchronous and Synchronous Collaboration and Online Awareness

MASE supports asynchronous collaboration by persisting to a database the state of any wiki pages one has worked on when he/she log out of MASE. This resembles the common practice in the real world where team members leave artifacts in a physical place for others to review or update when they work at different times.

MASE also supports synchronous work through its integration with the real-time collaboration tool, Microsoft NetMeeting. Every time when a team member logs into MASE, MASE tracks the network address of that team member's computer. Through the ListUser plug-in, MASE displays which members of a team are currently using the system, thus making all online team members aware of each other's presence. This is important for team members to establish informal and spontaneous communication with one another at ease.

4.5 Support for Agile Practices

As mentioned before, MASE supports agile practices through its library of plug-ins. For instance, project managers and customers can create iterations and user stories. MASE keeps track of all estimates made by the development team and suggests to both the development team and customers the appropriate size for the next iteration based on the developers' estimation accuracy from the previous iteration. Using the suggested iteration size, customers can prioritize user stories and move them from iteration to iteration or move them back to the product backlog. During the course of the project, both the customers and development team can track work progress at various granularities (project, iteration, user story) using the Whiteboard plug-in (see Figure 3) and view effort metrics for a particular individual or for the entire team.

Using the Whiteboard, developers can see the features and tasks allocated for each of the iterations in the project and track their time. Details of a user story or a task are stored in a wiki page allowing developers to annotate notes on them in free-formatted text. Leveraging MASE's integration with NetMeeting, developers can also perform distributed pair programming by sharing their code editor and collaborate on a design together using the shared whiteboard. Using the video and audio conferencing and multi-user text-chat features of NetMeeting, distributed team members who work at the same time can perform daily Scrum meetings.

Thus, MASE facilitates the following agile practices: release and iteration planning, distributed pair programming, collaborative design, and daily Scrum meetings.

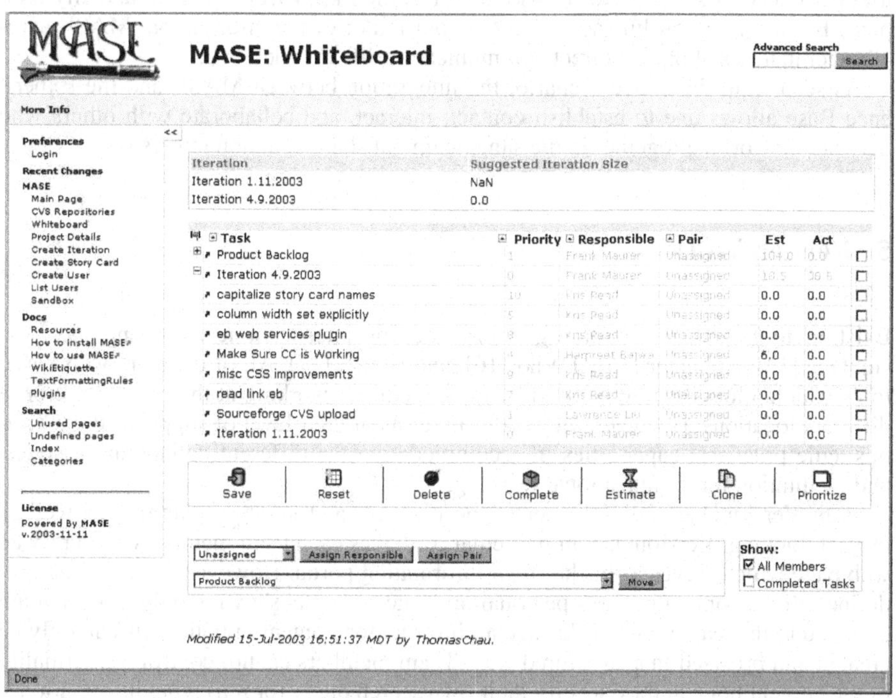

Fig. 3. MASE's project planning whiteboard

4.6 Facilitating Organizational Learning

To address the issue of organizational learning (or: inter-team learning), we adopt the view that workers learn and manage knowledge within communities of practice. We argue that facilitation for communities of practice is a critical part to support organizational learning [12]. MASE facilitates the establishment of communities of practice through its integration with the Experience Base, a tool that we have developed as an implementation of the Experience Factory concept [13] with a Wiki-like user interface similar to MASE. We illustrate MASE's support for communities of practice with the following example.

When Jill creates the task "Test shopping cart user interface" in MASE, she can associate the task with the process type "UI Testing". When Jill is ready to start working on that task, MASE will automatically create a wiki page for that task. Since the task is associated with the process type "UI Testing", Jill can embed the wiki page from the Experience Base that is dedicated to the topic "UI Testing" into the wiki page that contains details of the task she is working on. On the embedded "UI Testing" wiki page, Jill sees ideas contributed from Jack and Bill, whom she does not know and are from other teams in the company. Curious about their ideas and experiences, Jill posts her comments on the "UI Testing" wiki page.

Anecdotal evidence and thriving inter-organizational communities of practice that use Wiki servers suggest that this kind of informal knowledge sharing actually happens. By providing on-line access to the contributor of the information, MASE actually facilitates establishing direct communication Jill and Jack/Bill.

As seen from the above scenario, the integration between MASE and the Experience Base allows one to establish contact, interact, and collaborate with others who may not be working together in the same team but share common interests.

5 Related Work

Existing tools which support agile practices or inter-team learning include VersionOne [14], Xplanner [15], TWiki [16], and BORE [17]. All of them are web-based tools which allow them to be used by team members working at the same place or at different locations. However, they differ in terms of their level of support for the various agile practices, capabilities in accommodating the different collaboration styles, and facilitation for organizational learning.

Both VersionOne and Xplanner support release and iteration planning as well as project tracking. VersionOne, in particular, provides each team member with a private web page which serves as his/her own information portal. However, VersionOne predefines all the content in one's personal information portal showing only tasks that are assigned to the team member. In fact, all information content in both tools can only be created and browsed in a structured way. Team members cannot control the formality of the content nor can they specify their own search query for retrieving information.

TWiki also supports those agile team-related features provided by VersionOne and Xplanner but its usage is not targeted to agile development teams. It differentiates itself as a collaboration platform, not just a tool. This can be seen in the multitude of team-oriented tools it provides, such as event calendar, action tracker, drawing editor, and vote collection. As the name suggests, TWiki is developed based on the Wiki technology. Hence, TWiki and MASE share a lot in common: plug-in architecture, support for unstructured and structured information content, personal portal support, and full-text search. TWiki allows a set of web pages to be grouped together, known as a TWiki Web. This indirectly facilitates the establishment of communities of practice in that a community can have its own TWiki Web. One drawback of TWiki is that it provides no direct support for online team members to be aware of each other's presence. This limits the opportunities for team members to establish informal and spontaneous encounters with one another.

Unlike the other three tools, BORE does not directly support specific agile practices. It is an implementation of the Experience Factory concept. It provides an experience repository that contains experience collected from projects across the entire organization. These experiences are organized as cases, which are used to generate pre-defined tasks for a new project. A case is similar to a project task in nature. The generated set of tasks serves as a "best practice" guide. Project team members can diverge from the generated plan and not perform the suggested tasks if they deem the tasks to be inappropriate for the project situation at that time. In such cases, team members can submit their experiences and details of the tailored tasks in a structured format to the repository. A dedicated team of people is recommended to maintain the integrity of the cases stored in the repository. Despite its explicit support for inter-team learning, BORE does not provide the supports offered by the other three tools. Its repository-centric view of knowledge sharing also does not support the social context characteristic of the knowledge sharing culture in agile teams.

6 Concluding Remarks

Traditionally, software development teams follow the Tayloristic approach favoring division of labor, hence, the use of role-based teams. Role-based teams with hand-offs between job functions have the inherent problem of amplifying the problem of miscommunication due to indirect and long communication path. Agile development teams address this problem by using cross-functional teams which encourages direct communication and reduces the likelihood of miscommunication. They rely on approximative knowledge sharing by social interaction and fast feedback loops instead of structured (logical) representations. However, there are two major inherent limitations to the various knowledge sharing practices used by agile teams in their original forms. They support only co-located teams and they do not facilitate inter-team learning. In this paper, we describe a lightweight and integrated knowledge sharing environment, MASE, which facilitates agile software development team members to

- Collaborate as co-located and distributed team;
- To engage in synchronous and asynchronous work;
- And to collaborate with members from other teams in the company via communities of practice.

MASE is available under an open-source license (http://sern.ucalgary.ca/~milos) and is used by the MASE development team as well as in undergraduate and graduate courses in several institutions.

References

1. Highsmith III, J.A. (2000), *Adaptive Software Development: A Collaborative Approach to Managing Complex Systems,* Dorset House Publishing.
2. Beck, K (2000), *Extreme Programming Explained: Embrace Change,* Addison Wesley, Reading, MA.
3. Stapleton, J. (1997), *DSDM Dynamics System Development Method,* Addison Wesley, Reading, MA.
4. Beedle, M., Schwaber, K. (2001), *Agile Software Development with SCRUM,* Prentice Hall, Englewood Cliffs, NJ.
5. Cockburn, A. (2002), *Agile Software Development*; Addison Wesley, Reading, MA.
6. Ambler, S., Jeffries, R. (2002), *Agile Modeling: Effective Practice for Extreme Programming and the Unified Process,* Addison Wesley, Reading, MA.
7. Fitzpatrick G. (2001), Emergent Expertise Sharing in a New Community, in M.S. Ackerman, P. Volkmar & W. Volker, eds, *'Sharing Expertise: Beyond Knowledge Management'*, MIT Press, Cambridge, MA.
8. Haas, R., Aulbur, W., Thakar, S. (2000), Enabling Communities of Practice at EADS Airbus, in M.S. Ackerman, P. Volkmar & W. Volker, eds, *'Sharing Expertise: Beyond Knowledge Management'*, MIT Press, Cambridge, MA.
9. Greenburg, S., Roseman, M. (1998), Using a Room Metaphor to Ease Transitions in Groupware, in M.S. Ackerman, P. Volkmar & W. Volker, eds, *'Sharing Expertise: Beyond Knowledge Management'*, MIT Press, Cambridge, MA.
10. Cunningham, W., Leuf, B. (2001), *The Wiki Way Quick Collaboration on the Web,* Addison Wesley, Reading, MA.
11. JSPWiki http://www.jspwiki.org (Last Visited: September 25, 2003)
12. Erickson, T., Kellogg, W. (2001), Knowledge Communities: Online Environments for Supporting Knowledge Management and Its Social Context, in M.S. Ackerman, P. Volkmar & W. Volker, eds, *'Sharing Expertise: Beyond Knowledge Management'*, MIT Press, Cambridge, MA.
13. Basili, V., Caldiera, G., Romback, H. (1994), "Experience Factory", In *Encyclopedia of Software Engineering vol. 1,* J.J. Marciniak, Ed. John Wiley Sons.
14. VersionOne http://www.versionone.net (Last Visited: September 25, 2003)
15. Xplanner http://www.xplanner.org (Last Visited: September 25, 2003)
16. TWiki http://www.twiki.org (Last Visited: September 25, 2003)
17. Henninger, S., Ivaturi, A., Nuli, K., Thirunavukkaras, A. (2002), "Supporting Adaptable Methodologies to Meet Evolving Project Needs", in D. Wells, L. Williams, eds, *Proceedings of XP/Agile Universe 2002,* Springer, Berlin Heidelberg New York.

Logic and Approximation in Knowledge Based Systems

Michael M. Richter

Abstract. We consider approximation oriented and logic oriented representation of knowledge. The main focus is on problems where both representation methods have a natural place, such problem situations occur e.g. in e-commerce. The arising difficulty is that the inference method for both techniques are often not very compatible with each other. We make some suggestions for dealing with such problems. A major point here play the local-global principle and the concept of similarity. In particular, other concepts like utilities, probabilities, fuzzy sets, constraints and logical rules will be put in relation to similarities.

1 Introduction

About a decade ago the basic view on knowledge based systems was logic oriented: Some knowledge was declared in the knowledge base; this contained the assumptions from which conclusions were drawn. When the problem was presented as input to the system the solution was derived by logical reasoning. A particular form was logic programming and here this view was formulated in short by R. Kowalski: „Logic Programming = Logic + Control". For solving real world problems one performed a transformation of observations from the outside world into a formalism via some abstraction process. This process was often quite long, difficult and complicated but was considered as completed at some point. Essentially, this view was valid for most of knowledge based systems and their applications.
There were some undiscussed and tacitly assumed assumptions for this kind of problem solving:

- The knowledge base was stable during the inference process, in particular it was assumed to be relatively complete.
- Interactions with humans or the real world during the solution process were not expected.
- If uncertainty or vagueness was present then it was represented using some mathematically defined formalism like probability or numerical tolerances.

These assumptions were questioned when more and more complex problems have been tackled, in particular in socio-technical processes where humans and machines interact in various ways.
If these assumptions are missing three corresponding challenges arise:

- How to bring knowledge into the system when needed?
- How to organize and establish communication and cooperation between the different human and machine agents?

W. Lenski (Ed.): Logic versus Approximation, LNCS 3075, pp. 184–203, 2004.

- How to extend approximation techniques to symbolic domains and integrate them smoothly with logical reasoning methods?

We will consider two motivating scenarios.

Scenario 1:The first scenario deals with electronic commerce: A customer is searching for a certain product. In a first view this seems to be a logic oriented data base retrieval. This is no longer true if one allows vague, informally stated and incompletely formulated customer queries, and if the knowledge about the customer demand may change during the sales process. In addition, one wants to offer some product to the customer even if the ideally intended product is not available. Now the problem has become an optimization problem and we will discuss similarity based techniques as they have been developed in Case-Based Reasoning (CBR). We will consider even more involved situations were the products are represented in a compact form and the search has to consider logical inference operations as well.

Scenario 2:A second scenario comes from risk analysis. For portfolio it is often desiderable that not all assets go down drastically at the same time and a diversity is wanted for, which means that they are in a certain sense not similar. In order to formulate this, several methods have been developed in financial mathematics. They require, however, the presence of statistical data. For many symbolic information units they are unfortunately not available. As a common framework we again consider more generalized similarity based representations.

As a consequence of such and many related problems, the perspective on knowledge based systems was drastically extended. Incompleteness, imprecisement and changes in the assumptions as well as in the problem descriptions are not accidents anymore, they are rather considered as central to the whole area.

A general approach is to introduce approximation not only as a specific technique but also as a new paradigm. We will formulate this as

"Approximation Programming = Partial Orders + Control".

One of the advantages of logic was that it was theoretically well understood, especially concerning issues of semantics. For approximation on the other hand there are well understood foundations in real and functional analysis, probability theory etc. This is not yet quite so well established for approximation methods in symbolic domains occurring in AI applications.

For this purpose we will consider several areas where "inexact" concepts can be represented like utility functions, fuzzy sets, probabilities, and similarity measures. The reader is assumed to be familiar with these concepts, we will only elaborate certain aspects. Technically our main focus is the concept of similarity measures and a major aspect is to relate the discussed concepts to similarity. Similarity based methods have been developed in Case Based Reasoning to a fairly advanced level. Semantically similarity measures can be based on utilities and probabilities, syntactically fuzzy sets often are used as basic elements. These relations are discussed in 2.8.

On the logic side our main examples are rule applications and constraint propagation. We are in particular interested in problems where both, logical and approximation techniques are required. Experiences from practical applications as indicated above will provide illustrations of the difficulties and uses.

In the sequel we will first elaborate some relations and differences of the mentioned uncertainty concepts. On this basis we discuss problems arising when symbolic inferences are also present. This will be motivated and illustrated using the indicated application scenarios.

2 The Local-Global Principle

2.1 The Basic Concept

In order to relate different representation and derivation methods we need some structural principle that are shared by them and allows discovering common aspects as well as differences.

In order to formulate a systematic and general approach we introduce a structural representation principle.

The *local – global principle* for complex object description says:

1) There are *local* (atomic) description elements; for simplicity we assume that these are attributes.
2) Each object or concept A is *(globally)* described by some construction operator C from the local elements:

$$A = C(A_i \mid i \in I).$$

Here I is some index set for the atomic elements (i.e. the attributes).

The principle gives rise to two tasks:

a) The decomposition task: Break the object or concept down into atomic parts. This task happens often because the object may be presented globally and the parts are initially unknown.
b) The synthesis task: Compose an object or concept from simpler parts.

Both tasks play a role for relating the concepts we are interested in. For the objects under investigation the local – global principle has very different realizations. We will discuss it for representations of symbolic character as well as for approximation-oriented representations.

The claim is not that the principle itself is very innovative, in fact, it is quite standard. The point is the unified use of the principle in order to allow a systematic treatment of the different technique. For this purpose we will shortly introduce theses techniques and discuss them from this point of view.

2.2 Symbolic Representations

In symbolic representations the local-global principle is quite standard. Examples are component oriented descriptions of complex objects like machines or processes. They employ very often a part-of (or part-whole) hierarchy. The underlying language elements can be of different nature. Quite common are description logics; here we will restrict ourselves to attribute-value representations. There are various interpretations of the term "part-of"; for an overview see [8].

In order to describe relations between and classes of objects two important techniques are constraints and rules that are defined in terms of the underlying logical language, i.e. they refer to the atomic elements of the description. Both give rise to derivation methods, e.g. constraint propagation and rule chaining (forward and backward chaining).

2.3 Utility Functions and Preference Relations

Utility functions u in the first place operate on decisions or actions. These are, however, usually concerned with objects that are symbolically defined; e.g. one machine or one contract may be more useful than another one. Therefore we assume that the domain A of the utility function is a class of such objects. Utility functions assign real numbers as values to the elements of the domain:

$$u : A \rightarrow IR.$$

If u(a) > 0 we call it the benefit, otherwise the cost of a. Mostly u is considered as bounded; in this case we can assume without loss of generality that the values of u are in the interval [-1, 1].

The weaker relational formulation for utility functions uses preference relations. Both, utility functions and preference relations are usually complex. The local – global principle demands that they are defined on objects $A = C(A_i|i \in I)$.

The *local – global principle for utility functions* u (and analogously for preference relations) says that u can also be represented as

$$u(a) = G(u_i(a_i)|i \in I)$$

where the u_i are called the local utilities. It is desirable to model the situation in such a way that the global utility is a linear sum of local utilities; we will discuss this below:

$u(a) = u_w(a_1, ..., a_n)) = \Sigma (w_i \cdot u(a_i), 1 \leq i \leq n);$

the vector $w = (w_1, ..., w_n)$ of real valued coefficients is called the weight vector of the representation.

Often the utility function or rather the preference relation is known only globally. This means that one can relate two decisions but cannot give reasons for it because the local-global structure is unknown. The decomposition task then means mainly to identify the attributes, which determine the utility and can be influenced by the actor or decision maker and to determine the weight vector. The weight vector g reflects the degree of the influence of the attributes and its determination is the task of sensitivity analysis.

2.4 Fuzzy Sets

Fuzzy membership functions assign values to elements of an arbitrary domain U with respect to a certain property where one does not accept just "yes" or "no" assignments:

$$\mu : U \rightarrow [0, 1].$$

Membership functions μ are associated with *fuzzy subsets* (or predicates) P of U; the reference to P is denoted by μ_P. $\mu_P(a)$ is called the degree with which a has the property P. For many purposes one can assume that U is totally ordered and take for simplicity a real interval for U. There are two ways to combine different fuzzy sets:

1) Composition operators are mainly t-norms and co-t-norms (corresponding to conjunction and disjunction) or the different kinds of implications (e.g. Mamdami implication). With the implications one can describe conditional fuzzy degrees, as compared to conditional probabilities. These operators define a local – global principle for fuzzy membership functions defined on complex objects.
2) Fuzzy sets can be preconditions of different fuzzy – rules.

For our purposes two observations are important:

1) Norms and co-norms are symmetric, i.e. for complex fuzzy predicates there is no direct way to express importance or degree of influence for the constituents.
2) The membership functions identify all arguments with value 0; there are no negative degrees.

The way fuzzy logic deals with 2) is to define other fuzzy sets which cover the rest of the domain and occur in different fuzzy rules.

Fuzzy rules can be regarded as an integration of logic and approximation. This is done of the cost that the logic view is somehow no longer valid: The result of a rule application is no longer of a predicate and some defuzzyfication operation is needed.

2.5 Probabilities

We consider multivariate (n-dimensional) distribution functions H that are defined for vectors $(X_1, ..., X_n)$ of random variables from a viewpoint that is of interest in risk analysis. Their probability distributions also follow a *local – global principle.* The central concept for this is the notion of a *copula.*

Def.: An *n-dimensional copula* is a function $C: [0,1]^n \rightarrow [0,1]$ such that for all $a = (a_1, ..., a_n)$:

(i) if $a_i = 0$ for some i then $C(a) = 0$;

(ii) if $a_i = 1$ for all $i \neq j$ then $C(a) = a_j$;

(iii) if $a_i \leq b_i$ for all i then $0 \leq V_C([a, b])$ where $V_C([a, b])$ is the nth order difference on [a, b]: $V_C([a, b]) = \Delta^b{}_a C(t) = \Delta^{bn}{}_{an} ... \Delta^{b1}{}_{a1} C(t)$ and the first order differences are $\Delta^{bj}{}_{aj} C(t) = C(t_1, ..., t_{j-1}, b_j, t_{j+1}, ...t_n) - C(t_1, ..., t_{j-1}, a_j, t_{j+1}, ...t_n)$.

Condition (iii) is a monotonicity property.

The relevance of copulas is due to fact that they play the role of constructor functions which is explained by the following theorem.

Theorem (Sklar): For each n-dimensional distribution function H with margins $F_1, ..., F_n$ there is a copula C such that for all x

$$H(x_1, ..., x_n) = C(F_1(x_1), ..., F_n(x_n)).$$

Moreover, C is uniquely defined if all F_i are continuous. Conversely, if C is a copula, all F_i are distribution functions and H is defined by the above formula, then H is n-dimensional distribution function with margins $F_1, ..., F_n$.

Equivalently, copulas can be defined as n-dimensional distribution functions $C: [0, 1]^n \rightarrow [0, 1]^n$ with uniformly marginal distributions on [0, 1].

Another version of Sklar`s theorem is:

$$Prob(X_1 \leq x_1, ..., X_n \leq x_n) = C(F_1(x_1), ..., F_n(x_n)).$$

An important monotonicity property (see 2.7 below) is the following:

Theorem: If $(X_1, ..., X_n)$ is a vector of random variables with copula C and $(h_1, ..., h_n)$ are strictly increasing functions on the ranges $ran_i(X_i)$ then also $(h_1X_1, ..., h_nX_n)$ has copula C.

Related to copulas is the concept of *tail-dependency*, which is a form version of conditional distribution. Suppose $(X, Y)^T$ is a pair of continuous random variables with marginal distribution functions F and G.

Def.: (i) The coefficient of *upper tail dependency* is

$$\lambda_U = \lim(u\ 1)\ IP\{ Y > G^{-1}(u) \mid X > F^{-1}(u)\}$$

(provided that the limit exists).

(ii) If $\lambda_U > 0$ then X and Y are called *asymptotically dependent* in the upper tail, o they are otherwise *asymptotically independent* in the upper tail. These coefficients play a role in risk analysis.

For more details on copulas see [7] and [5].

2.6 Similarity Measures

In contrast to the above concepts, similarity functions are defined on pairs:

$$sim: U \times U \rightarrow [0, 1].$$

This notion can be extended to relate elements of different sets to each other:

$$sim: U \times V \rightarrow [0, 1].$$

Hence similarity functions can be considered as fuzzy sets of ordered pairs.

A kind of dualnotation for similarity measures are distance functions

$$d: U \times V \rightarrow [0, 1].$$

From principal point of view similarity measures and distance functions are equivalent.

This leads to several possible axioms for similarity measures:

1) $sim(x, x) = 1$ (reflexivity)
2) $sim(x, y) = sim(y, x)$ (symmetry).

For equality there were two more axioms:

3) $(x = y \wedge y = z) \rightarrow x = z$ (transitivity)
4) $(s(a) = t(a) \wedge a = b) \rightarrow (s(a) = t(b))$ (substitution axiom)

For distance functions an additional axiom is common:

5) $d(x, z) \leq d(x, y) + d(y, z)$ (triangle inequality).

Due to the changed view on similarity measures in the last years such axioms have been more or less given up. Originally the measure was concerned with the similarity of problem situations in order to apply previous experiences in some kind of analogical reasoning. The view on similarity between objects was that *they looked similar.* In this view similarity was thought as a form of fuzzy equality. From this point of view the transitivity axiom of equality was abandoned because small errors do add up.

This view was generalized extensively and the most recent extensions could run under the name „*partnership measure*". This leads to connecting similarity and utility as discussed below. Because the possible partners may come from different sets U and V the domain of sim had to be generalized to $U \times V$. The intention of partnership is that both objects cooperate more or less well as equal partners. In the extended view the measure compares things like *question and answer* or *demand and product.* As a consequence, the axioms of reflexivity and symmetry had to be given up. Also, the term "case base" from CBR is often replaced by expressions like product base, document base, etc.

An even more radical change came when similarity measures were regarded as a form of „*dependency measure*". Dependency introduces some kind of partial ordering, e.g. if sim(x, y) expresses the degree of dependency of y from x. A major point is that symmetry for sim is again no longer justified, i.e. sim(x, y) = sim(y, x) does not hold any more in general.

In such interpretations similarity is no longer coherent with the use of this term in everyday language. We still keep the term similarity not only because of historical reasons but because of the fact that the techniques developed for similarity reasoning apply here as well. Such techniques cover mainly retrieval algorithms, the structure of the database, the assessment of measures, and maintenance operations.

A natural question at this point asks whether similarity is still more than an arbitrary binary function and which general properties such measures might share. An answer is given by the application of the *local – global principle* to similarity measures:

There are measures sim_i on the domains of the A_i and there is some constructor function F such that $sim(a, b) = f(sim_i(a_i , b_i) \mid i \in I)$.

The sim_i are local measures (intended to reflect mainly domain properties) and sim is the global one (reflecting the utilities).

Common examples are the generalized Hamming measures with linear F defined by a measure vector $sim = (sim_1,\dots,sim_n)$ and a weight vector $w = (w_1, \dots,w_n)$ of non-negative coefficients:

$$H_{w,sim}((a_1, \dots, a_n), (b_1, \dots, b_n)) = \Sigma (w_i * sim_i(a_i , b_i) \mid 1 \le i \le n).$$

In the next session we will discover another general property of measures.

2.7 Monotonicity and Virtual Attributes

The concepts of uncertainties considered above are represented by numerical functions with real values. For all of them we can state a monotonicity property related to the local-global principle. For this we have to assume that all attributes A_i are equipped with a partial order $\le_i$.

Suppose that F is a utility function, a fuzzy membership function, a probability or a similarity function that is represented using the local-global principle from elementary objects.

Monotonicity axiom:

(i) For unary F: If $F(A) > F(B)$ then there is at least one $i \in I$ such that $F_i(A_i) > F(B_i)$;

(ii) For binary F: If $F(A, B) > F(A, C)$ then there is at least one $i \in I$ such that $F_i(A_i, B_i) > F_i(A_i, C_i)$.

The instance of the axiom for similarity measures reads as:

If $sim(A, B) > sim(A, C)$ then there is at least one $i \in I$ such that $sim_i(A_i , B_i) > sim_i(A_i , C_i)$.

This can be regarded as a partial order form of the substitution axiom (substituting equals by equals) for equalities. The importance of the axiom is twofold, it allows in general more efficient computations and it simplifies the assessment of utility functions and similarity measures.

Obviously generalized Hamming measures satisfy this axiom. The question arises whether the axiom can always be satisfied by a suitable choice of the measure. The answer depends on the underlying vocabulary. In fact, the axiom can sometimes not be satisfied if certain attributes are missing which depend certain relations between local attribute values. An example is the well-known XOR classification problem: Suppose the case base contains already three elements that are correctly classified, then there is no weighted Hamming measure for classifying the fourth element correctly using the nearest neighbor method.

A way out is to introduce additional (definable) attributes; in case of the XOR-problem the attribute XOR(x,y) will suffice. Such additional attributes are called *virtual attributes*. The purpose of introducing virtual attributes is usually to shift non-linear dependencies between attributes from the measure into the definition of virtual attributes.

In this view the monotonicity axiom is also a demand on the representation language.

2.8 Some Relations between the Concepts

If the relations between the concepts go beyond purely syntactic relations then we have to assume that the underlying intuitive meanings (i.e. the informal semantics) are related to each other. We can express this in different ways:

a) The induced relations of the functions (preference relations, similarity orderings) are compared; in particular their monotonicity behavior with respect to modified objects is studied.
b) The concepts are translated into each other.
c) Properties of the concepts and possible axioms for them are compared.

The relation between probabilities and similarity measures is postponed to section 3.4 because we encounter here both, numerical as well as symbolic attributes.

On the technical level the key for all comparisons is the assumption that all of the concepts follow a local – global principle.

2.8.1 Utilities and Fuzzy Sets

First we will relate utility functions and fuzzy membership functions. The classical semantic interpretation of the fuzzy membership treats the values as degrees with for the property in question. In this view fuzzy membership could be regarded as an undefined basic concept. Here we will relate it to decisions and utilities. The purpose of stating the degree of membership is to use it in subsequent actions as e.g. formulated in fuzzy control. These actions can again be performed with certain degrees. Another interpretation of the fuzzy values is with respect to utility functions:

$\mu(x) = 1$ means maximal utility; $\mu(x) < 1$ denotes the decrease of utility. Now we state the following axiom:

There are two directions to investigate:

a) To induce a utility function u by a membership function μ: u = ind(μ)
b) To induce a membership function μ by a measure μ: μ = ind*(u).

If a complex utility function is built from simpler local utilities, e.g. as a linear sum $u_g(a_1, ..., a_n)) = \Sigma (g_i \times u(a_i), 1 \leq i \leq n)$ then the $u(a_i)$ are often chosen as fuzzy membership functions. But instead of combining them by norms or co-norms here the weight vector is introduced which reflects the importance of the local entries. As mentioned, weights are not used in norms and co-norms. There is, however, another possibility for expressing different influences, namely to introduce several linguistic rules; we will not discuss this here.

2.8.2 Similarities and Fuzzy Sets

Similarity functions can syntactically be identified with fuzzy sets on ordered pairs. In order to define a similarity measure one needs not to start from pairs of objects. If we have simply a fuzzy set K of U and in addition a reference object x that satisfies $\mu_K(x) = 1$ then we can induce a measure by putting

$$sim(x, y) = \mu_K(y).$$

This induced measure satisfies also sim(x, x) = 1. Intuitively, sim(x, y) describes the way "x looks on U with respect to y". If there is a subset $P \subseteq U$ such that for each $x \in P$ we have some fuzzy subset K_x of U with membership functions $\mu_x(y)$, for which $\mu_K(x) = 1$ holds then we can again define a measure on U×P by

$$sim(y, x) = \mu_x(y), y \in U, x \in P.$$

Up to now similarity measures have only compared objects from U and V. There is no reason why this should not be extended to compare utility functions, probabilities and fuzzy membership functions. The latter is of particular interest in electronic commerce because demands as well as product descriptions often use vague concepts and the transformation into crisp attributes and predicates does not seem adequate.

We have the following possibilities:

1) sim(membership function, membership function)
2) sim(object, membership function)
3) sim(membership function, object)

Some examples:

1) Demand: I want a house near the university
   Offer: This house is in walking distance to the university
2) Demand: I want a house two kilometers from the university
   Offer: This house is still in walking distance to the university
3) Demand: I want a house in walking distance to the university
   Offer: This house is four kilometers from the university.

There are two known possibilities to compute the similarity of membership functions and μ_2 as seen in the next diagram:

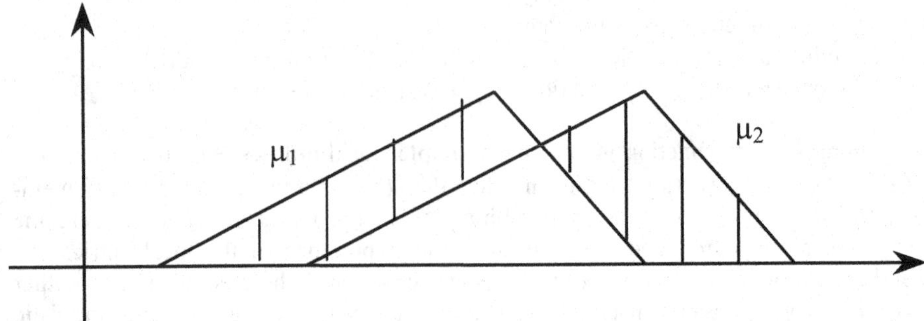

a) The *integral method* (uses the notion of distances):
Suppose F_i = Area between μ_i and the x - axis
d(m1, m2) = Size (F1Δ F2) (symmetric difference as indicated in the diagram);
b) The *crisp method* (uses again distances; it is furthermore assumed that the function attain their maximum at only one point):
Select a_i for which $\mu_i(a_i)$ maximal, i = 1,2; put $d(\mu_1, \mu_2) = | a1 - a2 |$ (the distance between the peaks in the diagram).

The disadvantage of the integral method is that two fuzzy functions with disjoint areas have always the same distance; the crisp method avoids this.
The disadvantage of the crisp method is that the shape of the curves does not play a role. The integral method avoids this.
A combined method is as follows:

If the areas are not disjoint apply the integral method.
If the areas are disjoint use the distance between the two points where both curves reach zero.

A generalization is obtained if the Euclidean distance $| a1 - a2 |$ is replaced by an arbitrary distance measure.

2.8.3 Utilities and Similarities

An obvious difference between the two concepts is that utilities have one and similarity measures have two arguments. Therefore we modify the range of similarity measures in order to compare their values with the values of a utility function:

$$\text{sim}: U \times V \rightarrow [-1, 1].$$

For this purpose we introduce for a measure sim:

$$f_{sim,x}(y) := \text{sim}(x,y).$$

If we interpret x = (a, sit) where a is an agent and sit is a (problem) situation then we denote the utility of a decision y for x by $u_x(y)$. In the equation

$$f_{sim,x}(y) = u_x(y)$$

either side can be taken in order to define the other side.
In a very general setting sim is defined on U × V where U contains problem and V contains decisions or actions. In this case sim(u, v) can be directly interpreted as the utility of v for u.

If we take utility functions as basic then they can provide also a meaning, i.e. a semantics to similarity in terms of utility. This means, the nearest neighbor of x is the decision y with the highest utility for a in the situation sit.
An equivalent possibility would have been to define $sim(x,y) := u_x(y)$.
Because utility functions are often not precisely known the equation $f_{sim,x}(y) = u_x(y)$ should be weakened. We propose that the following holds:
Assumption for the relation between similarity and utility (see [1]):
$f_{sim,x}$ and u_x are similarly ordered.
Here *similarly ordered* is defined as:
Def.: Two functions f g : X → IR are *similarly ordered* (or *concordant*) if

$$D(f, g, x, y) := (f(x) - f(y)) \times (g(x) - g(y)) \geq 0 \text{ for all } x, y \in X.$$

For the next considerations we allow to extend the range of measures to [-1, 1]
Examples:

a) U = queries, V = answers;
sim(q, a) > 0 means the answer has benefits, sim(q, a) < 0 means the answer creates costs.

b) U = demanded products, V = offered products;
sim(dp, op) > 0 means the offered product is acceptable, sim(dp, op) < 0 means the offered product is not acceptable.

c) U = V = products which can replace each other.
$sim(p_1, p_2) > 0$ means p_1 can replace p_2 to some degree while $sim(p_2, p_1) > 0$ says that p_1 can replace p_1 with this degree.

One intention is to apply the local-global principle to utility functions and similarity measures with the same constructor function. If we consider a generalized Hamming measure

$$H_{w,sim}((a_1, ..., a_n), (b_1, ..., b_n)) = \Sigma (w_i * sim_i(a_i, b_i) \mid 1 \leq i \leq n)$$

this means that it corresponds to a utility function

$$u_w(a_1, ..., a_n)) = \Sigma (w_i * u(a_i), 1 \leq i \leq n).$$

In a general situation the utility is not presented in a decomposed form and the similarity measure has to be found. This leads to a decomposition problem for utilities and a composition problem for measures.

3 Approximation and Logical Inference: Similarity, Constraints, and Rules

3.1 The General Problem

Approximation and optimization is always connected with one or more partial orders. On the other hand logical inference steps use terms like truth, falsehood or equality, there is no place for partial orders. In applications, however, we often encounter situations where both worlds play a role. In abstract setting a typical problem is of the following type:

Given a set A of objects, a set M of logical inference methods and a partial order ≤: Determine an optimal object O ∈ M(A) with respect to ≤, where M(A) is the deductive closure of A under methods M.
The major difficulty is that there is no a priori connection between the logical methods and the partial order, in particular the logical methods are in no way directed in the partial order sense. On the other hand, a coding of the problem in term s of real numbers and applying classical optimization techniques also provides difficulties because the symbolic origin of the problems usually does not admit assumptions that are needed by the optimization techniques.
We will now discuss this in more detail for some concrete examples.

3.2 Similarities and Rules

We consider a situation in electronic commerce as indicated above. A customer has a demand: A product with certain properties is wanted. The relation between the demand d and a product p is described by a similarity measure sim which means that we have to search for the nearest neighbor to d in space P of products (the product base).

Suppose now that the products are split into two sets:

- The products which are directly in some catalogue C of the product base;
- The products which can be obtained by adaptation from products in the catalogue.

The adaptation is performed by rules from a finite set R of rules and the set of all products P is the completion R(C) of C under the rules of R.
This means, the product base P is only partially represented explicitly. Such a product representation is quite common if the set of products is very large.
Formally the rules have an operator description. An operator is a partial function α: P → P, i.e., α transforms a product into a successor product. For systematic reasons we include the *trivial operator*, i.e. the identity, in the set of operators.
The rule representation of operators is in the form

$$R = (\text{preconditions}) \rightarrow (\text{actions}).$$

Concatenation of operator applications leads to a sequence $a = \alpha_1, \ldots, \alpha_n$ of operators which transforms a product p into an *adapted product* p_α, i.e.

$$p_\alpha = \alpha_n \circ \ldots \circ \alpha_1(p).$$

The set of operator sequences is denoted by OS.
In principle, there are two ways to make use of the rules, backward chaining and forward chaining.
In backward chaining one can asks queries to P which are answered by applying the rules backwards. An answer is positively given if the backward chaining leads to a product in C. This is, however, only successful if the wanted product is in fact an element of C and not only an approximation.
In case of forward chaining we assume that it is too complex to compute the application of R to C directly in order to obtain

- The nearest neighbor NN(d, p, C) to a demand d in the set C, i. e. without allowing rule applications.
- The nearest neighbor NN(d, p, P) in P, i.e. allowing rule applications.

The latter gives rise to a second type of similarity measures sim^{α} where α is some operator sequence:

$$sim^{\alpha}(q, p) := sim(q, p_{\alpha}).$$

In order to find p with NN(d, p, P) we need to compute $\max(sim^{\alpha}(d, p) \mid \alpha \in OS, p \in C)$. The problem is that the nearest neighbor search does not only investigate all products from C but also all operator sequences.

This search may not terminate because there may be arbitrary long sequences of operators. Even if this is not the case the method is in practical situations far too inefficient, in particular because the computations have to be done at run time. Therefore one is often satisfied with a good approximation of a nearest neighbor using computations that can be performed at compile time as much as possible. For this purpose knowledge about the operators is needed, in particular prediction knowledge how good the actions are. We will approach this problem stepwise.

3.2.1 Adaptation Knowledge

The underlying idea is to use *lazy rule applications* which means to apply operators as late as possible and to use knowledge about the result of the operator applications without executing them. As a consequence, the price to pay is that we will obtain only an approximation of the nearest neighbor in P. For this purpose we consider subsets $OS' \subseteq OS$ and allow only operator sequences from OS':

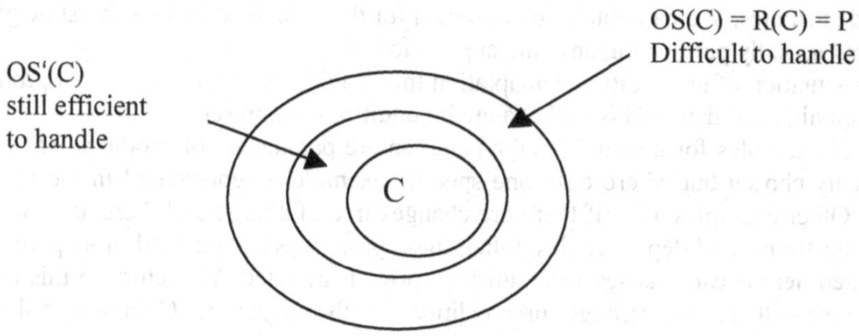

It follows directly:

If $OS'' \subseteq OS' \subseteq OS$ then $\max(sim^{\alpha}(d, p) \mid \alpha \in OS'', p \in C) \leq \max(sim^{\alpha}(d, p) \mid \alpha \in OS', p \in C) \leq \max(sim^{\alpha}(d, p) \mid \alpha \in OS, p \in C)$.- In order to fulfill them as much as possible we need knowledge about the rules which can be of different types:

Rule knowledge. The point is that such knowledge can often be used without actually knowing the definition of the rules and without applying them. It is useful to look at the attributes and knowledge about the possible changes of their values.

The rule knowledge consists in

- Knowledge about the applications of rules: When preconditions are satisfied and to which results the actions will lead.
- Preference rules for selecting rules.

a) Preconditions can be given in terms of attribute value restrictions (they lead also to a restriction of product types) and in terms of specific types of products
b) The actions must contain a list of affected attributes. In addition to the intended main effects there may be additional unavoidable effects.
c) Preference rules give orderings on the rules for operators that achieve some goal.

We will start with some simple subsets of OS and will then proceed to more powerful ones.

3.2.2 Unconditional Adaptation

These are the simplest rules but they provide a guideline for further treatment. *An unconditionally adaptable attribute* A has the properties

For any values x, y $\in$ dom(A) there is an operator sequence α such that

(i) $\alpha(x) = y$

(ii) No other attributes are affected by α.

The first condition says that the adaptation is not restricted by any constrains while the second is some independence property for attributes. Both conditions will have to be weakened in order to become applicable in practice.

We point out that simply the existence of such an α is of interest and not its specific form. The knowledge is used in the following way:

For each such attribute the local similarities sim_i between the and in value in the demand and in the intended the product are set to 1.

This gives rise to a new similarity measure sim* which has local values 1 at the selected attributes. We can obtain lower bound for the similarity to the nearest neighbor without actually performing any rule application.

In the situation of unconditional adaptation this is fairly is trivial because all products in P remain candidates. This will change if conditions are present.

Typical examples for unconditional adaptation are parameters of products which can be freely chosen but where only one specific example is represented in the product base. Other examples arise if there are changes free of charge and there are now further constraints and dependencies. Often these properties do not hold in its pure form as stated here because at least the attribute price is affected. We return to this below where we will consider stronger preconditions for the adaptation. Combinatorial difficulties arise if there are many dependencies involved.

3.2.3 Adaptation under Constraints

The principle arguments are the same as in the last section. We assume, however, not any more that the values of some attribute can be changed in an arbitrary way; the change is subject to constraints on the attributes.

We distinguish the following cases:

a) An operator is defined on the domain of some attribute A but there are constraints in the preconditions.
   If dom(A) is small then the possible changes can be enumerated explicitly. Otherwise one needs predicates on dom(A), e.g. an ordering $\leq$ on A. Typical constraints are e.g.

„if $y \leq x$ then α can change x to y"
„if $z \leq y \leq x$ then α can change x to y"
In these cases the local similarities can again be set to 1 or least increased which also increases the global similarity.

b) An add operator is defined for some component but there are constraints in the preconditions. The constraints can concern:
For some attributes A only certain values are allowed. Because these attributes may be the ones for which values are demanded in the query a local optimization has to take place in order to select the best available values or to approximate them sufficiently well. For remove operators this problem does not arise.
c) The adaptation is only possible for specific products. This is the first situation where actually products and not only their attribute values enter the scenario but elements of the product base too.

3.2.4 Adaptation When Dependencies Are Present

The dependencies we consider are twofold:

(i) The constraints for operator applications refer to other components and/or attributes as the one under concern.
(ii) The operator application has side effects that determine the values of other attributes or restrict these values.

Ad (i): In principle the preconditions in the rules concerning the desired operators have to be evaluated again. A serious combinatorial difficulty arises if there are several applications of operators that influence each other.

Ad (ii): We divide the set of involved attributes Z into two disjoint subsets, $Z = X \cup Y$ where X contains the attributes the values of which should be adapted and Y is the set of affected attributes. The partial weighted sum in the similarity measure concerning the predicates in Z is estimated by

- taking a lower bound for the increase in the local similarities with attributes in X
- taking an upper bound for the decrease in the local similarities with attributes in Y.

Again, for several dependencies this may become too inefficient.

3.2.5 Heuristics and Learning

As indicated, the situation becomes involved if many dependencies are present. Because the consequences of the dependencies are difficult to overlook one could make use of heuristics in order to represent adaptation knowledge. The question is how such heuristics can be obtained. A systematic approach has been developed by A.Stahl, (see [11] , [12] and [13]) who showed how machine learning methods could be used. In his approach genetic algorithms played the major role. Presently this seems to be the most promising approach.

3.3 Similarities and Constraints

We consider again a situation in electronic commerce with a very large set of products. Another way to obtain a compact description of the products is to introduce generalized products. Here products are grouped together by constraints on parameters describing them. This leads to the following definition:

Def.: (i) A *generalized object* is s set of objects

(ii) sim*(x, S) = max(sim(x, y) | y ∈ S)

The motivation is that the products in a generalized product allow a compact description, they are, however, in general not similar to each other.

Suppose the objects have only real valued attributes and the generalized sets are defined by linear or quadratic constraints. A demand x is a specific object and the computation of sim*(x, S) of x and a generalized object S leads to an optimization problem:

$$\max(\, sim(q, x) \mid h(x) \leq D)$$

where h is a set of m linear or quadratic functions and D is a vector of IR^m which define the constraints.

The attempt to use an arbitrary optimization method for this purpose has the following limitations:

1) The constraints defining S are in general not convex
2) The similarity measure is in general not differentiable.

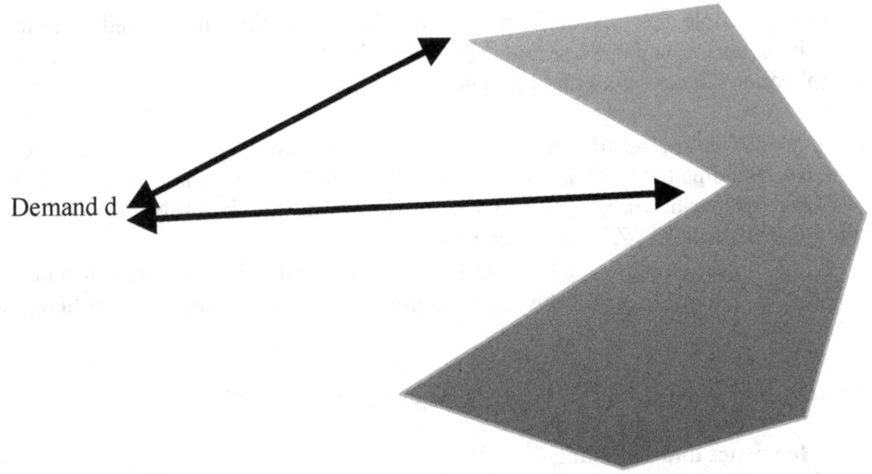

As an example we present a description of electronic switches (cf.[3]):

$$f \leq \begin{cases} -0.66 \cdot w + 115 & \text{if } s = 1 \\ -1.94 \cdot w + 118 & \text{if } s = 2 \\ -1.75 \cdot w + 88 & \text{if } s = 4 \\ -0.96 \cdot w + 54 & \text{if } s = 8 \\ -2.76 \cdot w + 57 & \text{if } s = no \end{cases} \quad a = \begin{cases} 1081 \cdot w^2 + 2885 \cdot w + 10064 & \text{if } s = 1 \\ 692 \cdot w^2 + 2436 \cdot w + 4367 & \text{if } s = 2 \\ 532 \cdot w^2 + 1676 \cdot w + 2794 & \text{if } s = 4 \\ 416 \cdot w^2 + 1594 \cdot w + 2413 & \text{if } s = 8 \\ 194 \cdot w^2 + 2076 \cdot w + 278 & \text{if } s = no \end{cases}$$

This is a very technical description of a Discrete Cosine Transformation which we will not investigate in detail; the meaning of the parameters is *f*: clock frequency, *a*: chip area, *w*: width (number of bits per input/output word) and *s*: subword (number of bits calculated per clock tick).
For solving the optimization problem several techniques have been presented in .[3] and .[9].
One method is to use Minkowski functionals (or gauges). For a compact set B a gauge with respect to B is defined as

$$\gamma_B(x) = \inf(\lambda \geq 0: x \in \lambda B).$$

For computing similarities the use of gauges is to compute a distance dist(x, y) from a point x to some y by putting x in the center of B and to enlarge or shrink B until it touches y; then one sets dist(x, y) = $\gamma_B(x - y)$. This technique can be used for non-convex generalized products to obtain upper and lower bounds for the similarity measures.

3.4 Similarities and Probabilities

The first connection between probabilities and similarities is that the semantics of the similarity measures sometimes can be reduced to probabilities. When similarity reasoning is used for classification problems a probabilistic semantics is very plausible: sim(a, b) = Prob(class(a) = class(b)).
This was discussed in [4]. Another approach uses evidence measures μ, see [10].
A second important relation is to consider similarities between random variables. A classical way to compare random variables is to use linear correlation. Let $(X, Y)^T$ be a vector of random variables.
Def.: The linear correlation coefficient for $(X, Y)^T$ is

$$\rho(X, Y) = \frac{Cov(X,Y)}{\sqrt{Var(X)} * \sqrt{Var(Y)}}$$

ρ(X, Y) is a similarity measure in the sense that it measures linear dependencies and it is quite natural for elliptical distributions. For other types of distributions the situation is, however, quite different.
The first extension provides a stochastic version of the concept "similarly ordered" given in 2.4.3:
Let (x^T, y^T) and $(\underline{x}^T, \underline{y}^T)$ be two observations of the continuous random variables (X^T, Y^T).
Def.: (x^T, y^T) and $(\underline{x}^T, \underline{y}^T)$ are called

- *concordant* if $(x - \underline{x})(y - \underline{y}) > 0$
- *discordant* if $(x - \underline{x})(y - \underline{y}) < 0$.

This has a probabilistic version (see [5]) where one considers independent vectors (X^T, Y^T) and $(\underline{X}^T, \underline{Y}^T)$ of continuous random variables. Then betweens these two vectors

- the probability of concordance is $Prob((X - \underline{X})(Y - \underline{Y}) > 0)$
- the probability of discordance is $Prob((X - \underline{X})(Y - \underline{Y}) < 0)$
- the order difference is $Q = Prob((X - \underline{X})(Y - \underline{Y}) > 0) - Prob((X - \underline{X})(Y - \underline{Y}) < 0)$.

This give rise to define a similarity measure called *measure of concordance* between random variables X and Y. The intention of concept of a copula introduced in 2.5 was to model dependencies between random variables, therefore it is no surprise that copulas are used for modeling concordance.
Let CRV denote the set of continuous random variables.
Def.: A mapping κ: CRV $\times$ CRV $\rightarrow$ [-1, 1] is a *measure of concordance* if

1) $\kappa(X, X) = 1$, $\kappa(X, -X) = -1$
2) $\kappa(X, Y) = k(Y, X)$
3) If X and Y are independent then $\kappa(X, Y) = 0$
4) $\kappa(X, -Y) = \kappa(-X, Y) = -\kappa(X, Y)$
5) If C(X, Y) and $\underline{C}(\underline{X}, \underline{Y})$ are the copulas corresponding to (X, Y) and ($\underline{X}$, $\underline{Y}$) and $C \leq \underline{C}$ then $\kappa(X, Y) \leq \kappa(\underline{X}, \underline{Y})$
6) If $\{(X_n, Y_n)\}$is a sequence with a copulas sequence $\{C_n\}$which converges pointwise to C(X, Y) then $\lim_n (\kappa(X_n, Y_n) = \kappa(X. Y)$.

The intentions are obvious: The value 1 means complete concordance and –1 means the opposite; 0 is a neutral value with no dependence. 5) is a monotonicity property while 6) is a technical requirement.
A simple example is the measure called *Kendall`s tau* (the order difference from above). Suppose ($\underline{X}$, $\underline{Y}$) is an independent copy of (X, Y).

Def.: $\tau(X, Y) = Q = \text{Prob}((X - \underline{X})(Y - \underline{Y}) > 0) - \text{Prob}((X - \underline{X})(Y - \underline{Y}) < 0)$.

Hence τ measures simply the difference of the probabilities for concordance and dis-concordance, which is intuitively clear for risk analysis if one assumes that X and Y represent costs.
The measure can be computed according to the following formula:

$$\tau(X, Y) = Q(C, \underline{C}) = 4\times \iint \underline{C}(u,v)dC(u,v) - 1.$$

where the integral is taken over $[0, 1]^2$. The factor 4 is due to the fact that the range of the measure is [-1, 1] instead of [0,1].

In risk analysis the statistical methods based on copulas and measures of concordance play an essential role. However, it was emphasized several times that qualitative, i.e. symbolic attributes are also important for describing risks. The problem is the lack of integration with numerical attributes. We suggest here that the similarity based approach provides a possibility for such an integration method.
For this we consider e.g. the Kendall`s τ measure. We consider it as a similarity measure and augment it with other local measures for qualitative attributes.
We consider some simple examples for such attributes in the context of investment. The risk to be avoided or minimized is that several assets of a portfolio go down drastically at the same time, i.e. that they are not concordant. We assume here that no distribution functions or statistical evidences for the relevancies of the values of the symbolic attributes are available. The similarity values arise rather from general experience and personal judgments and subjective probabilities of experts (see e.g. [6]).

Such experts will support in the first place similarity as a relation that can later on be refined to a similarity measure:

- Company type; range = {steel, military, energy, tourism, ...}. For the local similarity sim_{CT} with respect to concordancy it is reasonable to assume sim_{CT} (steel, tourism} < sim_{CT} (steel, military).
- Area; range = {EU, USA, Middle East, South America, ...}. For the local similarity sim_A with respect to concordancy it is reasonable to assume sim_A (EU, South America} < sim_A (EU, USA).

These attributes have to be associated with weights. The local measures and the weights reflect domain knowledge in the same way as distribution functions (if available) do.

The global measure has then two parts, one for qualitative and one for quantitative attributes. The point is that the measure handles them in a uniform way. This will be discussed in a future paper.

Another similarity measure with values in [0, 1] is the coefficient λ_U of upper tail dependence, which is relevant in risk management. This measure is not symmetric by its nature. Again, no qualitative information is yet integrated.

4 Summary

We have looked at some dependencies and relations between crisp concepts and methods in the sense of logic and several approaches that employ uncertainty and approximation. We have studied this for fuzzy sets, utility functions, probabilities and similarity measures; they are all connected with the concept of a partial order. In the definitorial sense fuzzy sets and probabilities are mostly basic, utilities and similarities use them in their definition and similarities are the last members in this chain.

The second point is of interest for the issue of the semantics of concepts. Here we regarded probabilities and utility functions as basic. Similarity is derived in the sense that its semantic is reduced to utility. The similarity is intended to describe an approximation of the primarily given but partially not exactly known utility function. Technically the approximation is essentially the computation of the nearest neighbor.

While there are many partial orderings there is only one concept of truth and logical deduction. The application of logical methods needs a step of severe abstraction as a prerequisite. The advantage of logical methods is exactness and preciseness. In many applications we encounter, however, a mixture of logical and approximation elements. This occurs in the statements of the problems and the assumptions as well as in the derivation steps. As typical examples some problems in e-commerce have been mentioned and some remarks on risk analysis have been added.

We have pointed out some difficulties that arise in such situations from the fact that often the techniques from logic and approximation are not compatible with each other. Some suggestions have been made to overcome the difficulties; an important one was the use of the local-global principle as a unifying element.

References

[1] Althoff, K.D., Richter, M.M. (99). Similarity and Utility. In: Mathematische Methoden der Wirtschaftswissenschaften Eds. Wolfgang Gaul and Martin Schader, Physica-Verlag, Heidelberg, 1999, pp. 403.413.

[2] Bergmann, R. Schmitt, S., Stahl,A.,. Vollrath, I. (2001). Utility-oriented matching: A new research direction for Case-Based Reasoning. In: Erfahrungen und Visionen. Proc. of the 1st Conference on Professional Knowledge Management. Shaker-Verlag 2001

[3] Bergmann, R., Moügouie, B. (2002). Similarity Assessment for Generalized Cases. Advances in Case-Based Reasoning, ed. Craw, S. and Preece, A. , Springer LNAI 2416. pp. 249-263.

[4] Faltings, B.(97). Probabilistic Indexing for Case-Based Prediction. Proc. ICCBR-97, Springer LNAI 1266 (1997).

[5] Embrecht, P.,. Lindskog, F., McNeill, A. (01). Modeling Dependence with Copulas and Applications to Risk Management. Preprint ETH Zurich 2001.

[6] Fishburn, P.C.(86). The Axioms of Subjective Probability. Statistical Science 1 (1986), p. 335-358.

[7] Joe, H.(97). Multivariate models and dependence concepts. Chapman and Hall 97.

[8] Lambrix, P. (2000). Part-Whole Reasoning in an Object –Centered Framework. LNAI 1771, 2000.

[9] Mouguoie, M., Richter, M.M. (2003). Generalized Cases, Similarity and Optimization.

[10] Richter, M. M.(95). On the Notion of Similarity in Case-Based Reasoning. In Mathematical and Statistical Methods in Artificial Intelligence (ed. G. della Riccia, R. Kruse, R. Viertl), Springer Verlag 171-184, 1995.

[11] Stahl, A., Gabel,T. [03). Using Evolution Programs to Learn Local Similarity Measures. Proc. Of the 5th Int. Conf. On CBR. Springer LNAI 2689, pp. 537-551.

[12] Stahl, A. (03). Defining Similarity Measures: Top-Down vs. Bottom-Up. Proc. of the 6th European Conference on Case-Based Reasoning, Springer LNAI 2416 2003, pp. 406-420.

[13] Stahl, A. (03). Approximation of Utility Functions by Learning Similarity Measures. This volume.

Author Index

Chau, Thomas 173

Hájek, Petr 1

Kämpke, Thomas 106
Klawonn, Frank 6
Kleine Büning, Hans 18
Kruse, Rudolf 6

Lenski, Wolfgang 77

Maurer, Frank 173
Mougouie, Babak 33

Oberschelp, Walter 43

Richter, Michael M. 184

Schinzel, Britta 59
Stahl, Armin 150
Stein, Benno 120

Wegener, Ingo 138

Zhao, Xishun 18